MRC Social & Public Health Sciences Unit
4 Lilybank Gardens
Glasgow G12 8RZ

SAS and R

Data Management, Statistical Analysis, and Graphics

SAS and R

Data Management, Statistical Analysis, and Graphics

Ken Kleinman

Nicholas J. Horton

CRC Press is an imprint of the
Taylor & Francis Group an **informa** business
A CHAPMAN & HALL BOOK

Chapman & Hall/CRC
Taylor & Francis Group
6000 Broken Sound Parkway NW, Suite 300
Boca Raton, FL 33487-2742

Printed in the United States of America on acid-free paper
10 9 8 7 6 5 4 3 2

International Standard Book Number: 978-1-4200-7057-6 (Hardback)

Library of Congress Cataloging-in-Publication Data

Kleinman, Ken.
SAS and R : data management, statistical analysis, and graphics / Ken Kleinman and Nicholas J. Horton.
p. cm.
Includes bibliographical references and index.
ISBN 978-1-4200-7057-6 (hard back : alk. paper)
1. SAS (Computer program language) 2. R (Computer program language) 3. SAS (Computer file) I. Horton, Nicholas J. II. SAS Institute. III. Title.

QA76.73.S27K54 2010
005.3--dc22 2009020819

Visit the Taylor & Francis Web site at
http://www.taylorandfrancis.com

and the CRC Press Web site at
http://www.crcpress.com

Contents

List of Figures

List of Tables

Preface

SAS™ (SAS Institute, 2009) and R (R development core team, 2009) are two statistical software packages used in many fields of research. SAS is commercial software developed by SAS Institute; it includes well-validated statistical algorithms. It can be licensed but not purchased. Paying for a license entitles the licensee to professional customer support. However, licensing is expensive and SAS sometimes incorporates new statistical methods only after a significant lag. In contrast, R is free, open-source software, developed by a large group of people, many of whom are volunteers. It has a large and growing user and developer base. Methodologists often release applications for general use in R shortly after they have been introduced into the literature. Professional customer support is not provided, though there are many resources for users. There are settings in which one of these useful tools is needed, and users who have spent many hours gaining expertise in the other often find it frustrating to make the transition.

We have written this book as a reference text for users of SAS and R. Our primary goal is to provide users with an easy way to learn how to perform an analytic task in both systems, without having to navigate through the extensive, idiosyncratic, and sometimes (often?) unwieldy documentation each provides. We expect the book to function in the same way that an English–French dictionary informs users of both the equivalent nouns and verbs in the two languages as well as the differences in grammar. We include many common tasks, including data management, descriptive summaries, inferential procedures, regression analysis, multivariate methods, and the creation of graphics. We also show some more complex applications. In toto, we hope that the text will allow easier mobility between systems for users of any statistical system.

We do not attempt to exhaustively detail all possible ways available to accomplish a given task in each system. Neither do we claim to provide the most elegant solution. We have tried to provide a simple approach that is easy to understand for a new user, and have supplied several solutions when it seems likely to be helpful. Carrying forward the analogy to an English-French dictionary, we suggest language that will communicate the point effectively, without listing every synonym or providing guidance on native idiom or eloquence.

Who should use this book

Those with an understanding of statistics at the level of multiple-regression analysis will find this book helpful. This group includes professional analysts who use statistical packages almost every day as well as statisticians, epidemiologists, economists, engineers, physicians, sociologists, and others engaged in research or data analysis. We anticipate that this tool will be particularly useful for sophisticated users, those with years of experience in only one system, who need or want to use the other system. However, intermediate-level analysts should reap the same benefit. In addition, the book will bolster the analytic abilities of a relatively new user of either system, by providing a concise reference manual and annotated examples executed in both packages.

Using the book

The book has three indices, in addition to the comprehensive table of contents. These include: 1) a detailed topic (subject) index in English; 2) a SAS index, organized by SAS syntax; and 3) an R index, describing R syntax. SAS users can use the SAS index to look up a task for which they know the SAS code and turn to a page with that code as well as the associated R code to carry out that task. R users can use the dictionary in an analogous fashion using the R index.

Extensive example analyses are presented; see Table C.1 (p. 277) for a comprehensive list. These employ a single dataset (from the HELP study), described in Appendix C. Readers are encouraged to download the dataset and code from the book website. The examples demonstrate the code in action and facilitate exploration by the reader.

Differences between SAS and R

SAS and R are so fundamentally distinct that an enumeration of their differences would be counter-productive. However, some differences are important for new users to bear in mind.

SAS includes data management tools that are primarily intended to prepare data for analysis. After preparation, analysis is performed in a distinct step, the implementation of which effectively cannot be changed by the user, though often extensive options are available. R is a programming environment tailored for data analysis. Data management and analysis are integrated. This means, for example, that calculating the BMI from weight and height can be treated as a function of the data, and as such is as likely to appear within a data analysis as in making a "new" piece of data to keep.

SAS Institute makes decisions about how to change the software or expand the scope of included analyses. These decisions are based on the needs of the user community and on corporate goals for profitability. For example, when changes are made, backwards-compatibility is almost always maintained, and documentation of exceptions is extensive. SAS Institute's corporate conservatism means that techniques are sometimes not included in SAS until they have been discussed in the peer-reviewed literature for many years. While the R Core Team controls base functionality, a very large number of users have developed functions for R. Methodologists often release R functions to implement their work concurrently with publication. While this provides great flexibility, it comes at some cost. A user-contributed function may implement a desired methodology, but code quality may be unknown, documentation scarce, and paid support nonexistent. Sometimes a function which once worked may become defunct due to a lack of backwards-compatibility and/or the author's inability to, or lack of interest in, updating it.

Other differences between SAS and R are worth noting. Data management in SAS is undertaken using row by row (observation-level) operations. R is inherently a vector-based language, where columns (variables) are manipulated. R is case-sensitive, while SAS is generally not.

Where to begin

We do not anticipate that the book will be read cover to cover. Instead, we hope that the extensive indexing, cross-referencing, and worked examples will make it possible for readers to directly find and then implement what they need. A user new to either SAS or R should begin by reading the appropriate Appendix for that software package, which includes a sample session and overview.

On the web

The book website at `http://www.math.smith.edu/sasr` includes the table of contents, the indices, the HELP dataset, example code in SAS and R, and a list of erratum.

Acknowledgments

We would like to thank Rob Calver, Shashi Kumar, and Sarah Morris for their support and guidance at Informa CRC/Chapman and Hall, the Department of Statistics at the University of Auckland for graciously hosting NH during a sabbatical leave, and the Office of the Provost at Smith College. We also thank Allyson Abrams, Tanya Hakim, Ross Ihaka, Albyn Jones, Russell Lenth, Brian McArdle, Paul Murrell, Alastair Scott, David Schoenfeld, Duncan Temple Lang, Kristin Tyler, Chris Wild, and Alan Zaslavsky for contributions to SAS, R, or LaTeX programming efforts, comments, guidance and/or helpful suggestions on drafts of the manuscript.

Above all we greatly appreciate Sara and Julia as well as Abby, Alana, Kinari, and Sam, for their patience and support.

Amherst, MA and Northampton, MA
March, 2009

Chapter 1

Data management

This chapter reviews basic data management, beginning with accessing external datasets, such as those stored in spreadsheets, ASCII files, or foreign formats. Important tasks such as creating datasets and manipulating variables are discussed in detail. In addition, key mathematical, statistical, and probability functions are introduced.

1.1 Input

Both SAS and R provide comprehensive support for data input and output. In this section we address aspects of these tasks.

SAS native datasets are rectangular files with data stored in a special format. They have the form `filename.sas7bdat` or something similar, depending on version. In the following, we assume that files are stored in directories and that the locations of the directories in the operating system can be labeled using Windows syntax (though SAS allows UNIX/Linux/Mac OS X-style forward slash as a directory delimiter on Windows). Other operating systems will use local idioms in describing locations.

R organizes data in dataframes, or connected series of rectangular arrays, which can be saved as platform independent objects. R also allows UNIX-style directory delimiters (forward slash) on Windows.

1.1.1 Native dataset

HELP example: see 4.6

SAS

```
libname libref "dir_location";
data ds;
   set libref.sasfilename;    /* Note: no file extension */
   ...
run;
```

or

```
data ds;
   set "dir_location\sasfilename.sas7bdat"; /* Windows only */
   set "dir_location/sasfilename.sas7bdat";
                         /* works on all OS including Windows */
   ...
run;
```

Note: The file `sasfilename.sas7bdat` is created by using a `libref` in a `data` statement; see 1.2.1.

R

```
load(file="dir_location/savedfile")    # works on all OS including Windows
load(file="dir_location\\savedfile")   # Windows only
```

Note: Forward slash is supported as a directory delimiter on all operating systems; a double backslash is supported under Windows. The file `savedfile` is created by `save()` (see 1.2.1).

1.1.2 Fixed format text files

See also 1.1.3 (read more complex fixed files) and 6.4 (read variable format files)

SAS

```
data ds;
   infile 'C:\file_location\filename.ext';
   input varname1 ... varnamek;
run;
```

or

```
filename filehandle 'file_location/filename.ext';

proc import datafile=filehandle
   out=ds dbms=dlm;
   getnames=yes;
run;
```

Note: The `infile` approach allows the user to limit the number of rows read from the data file using the `obs` option. Character variables are noted with a trailing '$', e.g., use a statement such as `input varname1 varname2 $ varname3` if the second position contains a character variable (see 1.1.3 for examples). The `input` statement allows many options and can be used to read files with variable format (6.4.1).

In `proc import`, the `getnames=yes` statement is used if the first row of the input file contains variable names (the variable types are detected from the data). If the first row does not contain variable names then the `getnames=no` option should be specified. The `guessingrows` option (not shown) will base the variable formats on other than the default 20 rows. The `proc import` statement will accept an explicit file location rather than a file associated by the `filename` statement as in section 4.6.

Note that in Windows installations, SAS accepts either slashes or backslashes to denote directory structures. For Linux, only forward slashes are allowed. Behavior in other operating systems may vary.

In addition to these methods, files can be read by selecting the `Import Data` option on the `file` menu in the GUI.

R

```
ds <- read.table("dir_location\\file.txt", header=TRUE) # Windows only
```

or

```
ds <- read.table("dir_location/file.txt", header=TRUE) # all OS (including
                                                       # Windows)
```

Note: Forward slash is supported as a directory delimiter on all operating systems; a double backslash is supported under Windows. If the first row of the file includes the name of the variables, these entries will be used to create appropriate names (reserved characters such as '$' or '[' are changed to '.') for each of the columns in the dataset. If the first row doesn't include the names, the `header` option can be left off (or set to `FALSE`), and the variables will be called `V1, V2, ... Vn`. A limit on the number of lines to be read can be specified through the `nrows` option. The `read.table()` function can support reading from an URL as a filename (see also 1.1.6) or browse files interactively using `file.choose()` (see 1.7.5).

1.1.3 Reading more complex text files

See also 1.1.2 (read fixed files) and 6.4 (read variable format files)

Text data files often contain data in special formats. One common example is date variables. Special values can be read in using informats (A.6.4). As an example below we consider the following data.

```
1 AGKE 08/03/1999 $10.49
2 SBKE 12/18/2002 $11.00
3 SEKK 10/23/1995 $5.00
```

SAS

```
data ds;
   infile 'C:\file_location\filename.dat';
   input id initials $ datevar mmddyy10. cost dollar7.4;
run;
```

Note: The SAS informats (A.6.4) denoted by the `mmddyy10.` and `dollar7.4` inform the `input` statement that the third and fourth variables have special forms and should not be treated as numbers or letters, but read and interpreted according to the rules specified. In the case of `datevar`, SAS reads the date appropriately and stores a SAS date value (section A.6.4). For `cost`, SAS ignores the '$' in the data and would also ignore commas, if they were present. The `input` statement allows many options for additional data formats and can be used to read files with variable format (6.4.1).

Other common features of text data files include very long lines and missing data. These are addressed through the `infile` or `filename` statements. Missing data may require the `missover` option to the `infile` statement as well as listing the columns in which variables appear in the dataset in the `input` statement. Long lines (many columns in the data file) may require the `lrecl` option to the `infile` or `filename` statement. For a thorough discussion, see the on-line help: Contents; SAS Products; Base SAS; SAS 9.2 Language Reference: Concepts; DATA Step Concepts; Reading Raw Data; Reading Raw Data with the INPUT statement.

R

```
tmpds <- read.table("file_location/filename.dat")
id <- tmpds$V1
initials <- tmpds$V2
datevar <- as.Date(as.character(tmpds$V3), "%m/%d/%y")
cost <- as.numeric(substr(tmpds$V4, 2, 100))
ds <- data.frame(id, initials, datevar, cost)
rm(tmpds, id, initials, datevar, cost)
```

Note: In R, this task is accomplished by first reading the dataset (with default names from `read.table()` denoted `V1` through `V4`). These objects can be manipulated using `as.character()` to undo the default coding as factor variables, and coerced to the appropriate data types. For the `cost` variable, the dollar signs are removed using the `substr()` function. Finally, the individual variables are gathered together as a dataframe.

1.1.4 Comma separated value (CSV) files

HELP example: see 1.13.1

SAS

```
data ds;
   infile 'dir_location\filename.csv' delimiter=',';
   input varname1 ... varnamek;
run;
```

or

```
proc import datafile='dir_location\full_filename'
   out=ds dbms=dlm;
   delimiter=',';
   getnames=yes;
run;
```

Note: Character variables are noted with a trailing '$', e.g., use a statement such as `input varname1 varname2 $ varname3` if the second column contains characters. The `proc import` syntax allows for the first row of the input file to contain variable names, with variable types detected from the data. If the first row does not contain variable names then use `getnames=no`.

In addition to these methods, files can be read by selecting the `Import Data` option on the `file` menu in the GUI.

R

```
ds <- read.csv("dir_location/file.csv")
```

Note: A limit on the number of lines to be read can be specified through the `nrows` option. The command `read.csv(file.choose())` can be used to browse files interactively (see section 1.7.5). The comma-separated file can be given as an URL (see 1.1.6).

1.1.5 Reading datasets in other formats

HELP example: see 3.7.1

SAS

```
libname ref spss 'filename.sav';   /* SPSS */
libname ref bmdp 'filename.dat';   /* BMDP */
libname ref v6 'filename.ssd01;     /* SAS vers. 6 */
libname ref xport 'filename.xpt'; /* SAS export */
libname ref xml 'filename.xml';    /* XML */

data ds;
set ref.filename;
run;
```

or

```
proc import datafile="filename.ext' out=ds
   dbms=excel;                                   /* Excel */
run;

... dbms=access; ...                             /* Access */
... dbms=dta; ...                                /* Stata */
```

Note: The `libname` statements above refer to files, rather than directories. The extensions shown above are those typically used for these file types, but in any event the full name of the file, including the extension, is needed in the `libname` statement. In contrast, only the file name (without the extension) is used in the `set` statement. The data type options specified above in the `libname` statement and `dbms` option are available in Windows. To see what's available under other operating systems, check in the on-line help: Contents, Using SAS in Your Operating Environment, SAS 9.2 Companion for <your OS>, Features of the SAS language for <your OS>, Statements under <your OS>, Libname statement.

In addition to these methods, files can be read by selecting the `Import Data` option on the `file` menu in the GUI.

R

```
library(foreign)
ds <- read.dbf("filename.dbf")            # DBase
ds <- read.epiinfo("filename.epiinfo") # Epi Info
ds <- read.mtp("filename.mtp")            # Minitab portable worksheet
ds <- read.octave("filename.octave")     # Octave
ds <- read.ssd("filename.ssd")            # SAS version 6
ds <- read.xport("filename.xport")       # SAS XPORT file
ds <- read.spss("filename.sav")           # SPSS
ds <- read.dta("filename.dta")            # Stata
ds <- read.systat("filename.sys")        # Systat
```

Note: The `foreign` library can read Stata, Epi Info, Minitab, Octave, SPSS, and Systat files (with the caveat that SAS files may be platform dependent). The `read.ssd()` function will only work if SAS is installed on the local machine.

1.1.6 URL

SAS

HELP example: see 2.6.1

```
filename urlhandle url 'http://www.math.smith.edu/sasr/testdata';

filename urlhandle url 'http://www.math.smith.edu/sasr/testdata'
   user='your_username' pass='your_password';

proc import datafile=urlhandle out=ds dbms=dlm;
run;
```

Note: The latter `filename` statement is needed only if the URL requires a username and password. The `urlhandle` used in a `filename` statement can be no longer than 8 characters. A `urlhandle` can be used in an `import` procedure as shown, or with an `infile` statement in a `data` step (see 6.4). The `import` procedure supports many file types through the `dbms` option; `dbms=dlm` without the `delimiter` option (section 1.1.4) is for space-delimited files.

R

```
urlhandle <- url("http://www.math.smith.edu/sasr/testdata")
ds <- readLines(urlhandle)
```

or

```
ds <- read.table("http://www.math.smith.edu/sasr/testdata")
```

or

```
ds <- read.csv("http://www.math.smith.edu/sasr/file.csv")
```

Note: The `readLines()` function reads arbitrary text, while `read.table()` can be used to read a file with cases corresponding to lines and variables to fields in the file (the `header` option sets variable names to entries in the first line). The `read.csv()` function can be used to read comma separated values. Access through proxy servers as well as specification of username and passwords is provided by the function `download.file()`. A limit on the number of lines to be read can be specified through the `nrows` option.

1.1.7 XML (extensible markup language)

A sample (flat) XML form of the HELP dataset can be found at `http://www.math.smith.edu/sasr/datasets/help.xml`. The first ten lines of the file consist of:

```
?xml version="1.0" encoding="iso-8859-1" ?>
<TABLE>
   <HELP>
      <id> 1 </id>
      <e2b1 Missing="." />
      <g1b1> 0 </g1b1>
      <i11 Missing="." />
      <pcs1> 54.2258263 </pcs1>
      <mcs1> 52.2347984 </mcs1>
      <cesd1> 7 </cesd1>
```

Here we consider reading simple files of this form. While support is available for reading more complex types of XML files, these typically require considerable additional sophistication.

SAS

```
libname ref xml 'dir_location\filename.xml';

data ds;
   set ref.filename_without_extension;
run;
```

Note: The `libname` statement above refers to a file name, rather than a directory name. The "xml" extension is typically used for this file type, but in any event the full name of the file, including the extension, is needed.

R

```
library(XML)
urlstring <- "http://www.math.smith.edu/sasr/datasets/help.xml"
doc <- xmlRoot(xmlTreeParse(urlstring))
tmp <- xmlSApply(doc, function(x) xmlSApply(x, xmlValue))
ds <- t(tmp)[,-1]
```

Note: The XML library provides support for reading XML files. The xmlRoot() function opens a connection to the file, while xmlSApply() and xmlValue() are called recursively to process the file. The returned object is a character matrix with columns corresponding to observations and rows corresponding to variables, which in this example are then transposed.

1.1.8 Data entry

SAS

```
data ds;
input x1 x2;
cards;
1 2
1 3
1.4 2
123 4.5
;
run;
```

Note: The above code demonstrates reading data into a SAS dataset within a SAS program. The semicolon following the data terminates the data step, meaning that a run statement is not actually required. The input statement used above employs the syntax discussed in 1.1.2. In addition to this option for entering data within SAS, there is a GUI-based data entry/editing tool called the Table Editor. It can be accessed using the mouse through the Tools menu, or by using the viewtable command on the SAS command line.

R

```
x <- numeric(10)
data.entry(x)
```

or

```
x1 <- c(1, 1, 1.4, 123)
x2 <- c(2, 3, 2, 4.5)
```

Note: The data.entry() function invokes a spreadsheet that can be used to edit or otherwise change a vector or dataframe. In this example, an empty numeric vector of length 10 is created to be populated. The data.entry() function differs from the edit() function, which leaves the objects given as argument unchanged, returning a new object with the desired edits (see also fix()).

1.2 Output

1.2.1 Save a native dataset

SAS *HELP example:* see 1.13.1

```
libname libref "dir_location";

data libref.sasfilename;
set ds;
run;
```

Note: A SAS dataset can be read back into SAS using a set statement with a libref, see 1.1.1.

R

```
save(robject, file="savedfile")
```

Note: An object (typically a dataframe, or a list of objects) can be read back into R using `load()` (see 1.1.1).

1.2.2 Creating files for use by other packages

See also 1.2.7 (write XML) *HELP example:* see 1.13.1

SAS

```
libname ref spss 'filename.sav';  /* SPSS */
libname ref bmdp 'filename.dat';  /* BMDP */
libname ref v6 'filename.ssd01';  /* SAS version 6 */
libname ref xport 'filename.xpt'; /* SAS export */
libname ref xml 'filename.xml';   /* XML */

data ref.filename_without_extension;
set ds;
```

or

```
proc export data=ds outfile='file_location_and_name'
   dbms=csv;            /* comma-separated values */

...dbms=dbf;            /* dbase 5,IV,III */
...dbms=excel;          /* excel */
...dbms=tab;            /* tab-separated values */
...dmbs=access;         /* Access table */
...dbms=dlm;            /* arbitrary delimiter; default is space,
                           others with delimiter=char statement */
```

Note: The `libname` statements above refer to file names, rather than directory names. The extensions shown above are those conventionally used but the option specification determines the file type that is created.

R

```
library(foreign)
write.dta(ds, "filename.dta")
write.dbf(ds, "filename.dbf")
write.foreign(ds, "filename.dat", "filename.sas", package="SAS")
```

Note: Support for writing dataframes in R is provided in the `foreign` library. It is possible to write files directly in Stata format (see `write.dta()`) or DBF format (see `write.dbf()` or create files with fixed fields as well as the code to read the file from within Stata, SAS, or SPSS using `write.foreign()`).

As an example with a dataset with two numeric variables X_1 and X_2, the call to `write.foreign()` creates one file with the data and the SAS command file `filename.sas`, with the following contents.

```
data ds;
   infile "file.dat" dsd lrecl=79;
   input x1 x2;
run;
```

This code uses `proc format` (1.4.12) statements in SAS to store string (character) variables. Similar code is created for SPSS using `write.foreign()` with appropriate `package` option.

1.2.3 Creating datasets in text format

SAS

```
proc export data=ds outfile='file_location_and_name'
   dbms=csv;              /* comma-separated values */

...dbms=tab;              /* tab-separated values */
...dbms=dlm;              /* arbitrary delimiter; default is space,
                             others with delimiter= statement */
```

R

```
library(foreign)
write.csv(ds, file="full_file_location_and_name")
```

or

```
library(foreign)
write.table(ds, file="full_file_location_and_name")
```

Note: The `sep` option to `write.table()` can be used to change the default delimiter (space) to an arbitrary value.

1.2.4 Displaying data

HELP example: see 3.7.2

See also 1.3.3 (values of variables in a dataset)

SAS

```
title1 'Display of variables';
footnote1 'A footnote';
proc print data=ds;
   var x1 x3 xk x2;
   format x3 dollar10.2;
run;
```

Note: For `proc print` the `var` statement selects variables to be included. The format statement, as demonstrated, can alter the appearance of the data; here x_3 is displayed as a dollar amount with 10 total digits, two of them to the right of the decimal. The keyword `_numeric_` can replace the variable name and will cause all of the numerical variables to be displayed in the same format. See section A.6.4 for further discussion.

See sections A.6.3, A.6.2, and A.6.1 for ways to limit which observations are displayed. The `var` statement, as demonstrated, can cause the variables to be displayed in the desired order. The `title` and `footnote` statements and related statements `title1`, `footnote2`, etc. allow headers and footers to be added to each output page. Specifying the command with no argument will remove the title or footnote from subsequent output.

SAS also provides `proc report` and `proc tabulate` to create more customized output.

R

```
dollarcents <- function(x)
   return(paste("$", format(round(x*100, 0)/100, nsmall=2), sep=""))
data.frame(x1, dollarcents(x3), xk, x2)
```

or

```
ds[,c("x1", "x3", "xk", "x2"]
```

Note: A function can be defined to format a vector as U.S. dollar and cents by using the `round()` function (see 1.8.4) to control the number of digits (2) to the right of the decimal. Alternatively, named variables from a dataframe can be printed. The `cat()` function can be used to concatenate values and display them on the console (or route them to a file using the `file` option). More control on the appearance of printed values is available through use of `format()` (control of digits and justification), `sprintf()` (use of C-style string formatting) and `prettyNum()` (another routine to format using C-style specifications).

1.2.5 Number of digits to display

HELP example: see 1.13.1

SAS lacks an option to control how many significant digits are displayed in procedure output, in general (an exception is `proc means`). For reporting purposes, one should save the output as a dataset using `ODS`, then use the `format` statement (1.2.4, A.6.4) with `proc print` to display the desired precision as demonstrated in section 3.7.2.

R

```
options(digits=n)
```

Note: The `options(digits=n)` command can be used to change the default number of decimal places to display in subsequent R output. To affect the actual significant digits in the data, use the `round()` function (see 1.8.4).

1.2.6 Creating HTML formatted output

SAS

```
ods html file="filename.html";
...
ods html close;
```

Note: Any output generated between an `ods html` statement and an `ods hmtl close` statement will be included in an HTML (hyper-text markup language) file (A.7.2). By default this will be displayed in an internal SAS window; the optional `file` option shown above will cause the output to be saved as a file.

R

```
library(prettyR)
htmlize("script.R", title="mytitle", echo=TRUE)
```

Note: The `htmlize()` function within `library(prettyR)` can be used to produce HTML (hypertext markup language) from a script file (see B.2.1). The `cat()` function is used inside the script file (here denoted by `script.R`) to generate output. The `hwriter` library also supports writing R objects in HTML format.

1.2.7 Creating XML datasets and output

In R, the XML library provides support for writing XML files (see also 1.1.5, write foreign files and Further resources).

SAS

```
libname ref xml 'dir_location\filename.xml';

data ref.filename_without_extension;
   set ds;
run;
```

or

```
ods docbook file='filename.xml';
...
ods close;
```

Note: The libname statement can be used to write a SAS dataset to an XML-formatted file. It refers to a file name, rather than a directory name. The file extension xml is conventionally used but the xml specification, rather than the file extension, determines the file type that is created.

The ods docbook statement, in contrast, can be used to generate an XML file displaying procedure output; the file is formatted according to the OASIS DocBook DTD (document type definition).

1.3 Structure and meta-data

1.3.1 Access variables from a dataset

HELP example: see 1.13.1

In SAS, every data step or procedure refers to a dataset explicitly or implicitly. Any variable in that dataset is available without further reference. In R, variable references must contain the name of the object which includes the variable, unless the object is attached, see below.

R

```
attach(ds)
detach(ds)
with(ds, mean(x))
```

Note: The command attach() will make the variables within the named dataset available in the workspace (otherwise they need to be accessed using the syntax ds$var1 unless they are in the workspace). The detach() function removes them from the workspace (and is recommended when the local version is no longer needed, to avoid name conflicts). The with() and within() functions provide another way to access variables within a dataframe without having to worry about later detaching the dataframe. Many functions (e.g., lm()) allow specification of a dataset to be accessed using the data option.

The detach() function is also used to remove a package from the workspace: more information can be found in section B.4.5. This is sometimes needed if a package overrides a built-in function within R. As an example the command detach("package:packagename") will detach a package that had been loaded using library(packagename).

1.3.2 Names of variables and their types

HELP example: see 1.13.1

SAS

```
proc contents data=ds;
run;
```

R

```
str(ds)
```

Note: The command `sapply(ds, class)` will return the names and classes (e.g., numeric, integer or character) of each variable within a dataframe, while running `summary(ds)` will provide an overview of the distribution of each column.

1.3.3 Values of variables in a dataset

HELP example: see 1.13.2

SAS

```
proc print data=ds (obs=nrows);
   var x1 ... xk;
run;
```

Note: The integer `nrows` for the `obs=nrows` option specifies how many rows to display, while the `var` statement selects variables to be displayed (A.6.1). Omitting the `obs=nrows` option or `var` statement will cause all rows and all variables in the dataset to be displayed, respectively.

R

```
print(ds)
```

or

```
View(ds)
```

or

```
edit(ds)
```

or

```
ds[1:10,]
ds[,2:3]
```

Note: The `print()` function lists the contents of the dataframe (or any other object), while the `View()` function opens a navigable window with a read-only view. The contents can be changed using the `edit()` function. Alternatively, any subset of the dataframe can be displayed on the screen using indexing, as in the final example. In the first example, the first 10 records are displayed, while in the second, the second and third variables. Variables can also be specified by name using a character vector index (see B.4.2). The `head()` function can be used to display the first (or last) values of a vector, dataset, or other object.

1.3.4 Rename variables in a dataset

SAS

```
data ds2;
set ds (rename = (old1=new1 old2=new2 ...));
...
```

or

```
data ds;
...
rename old=new;
```

R

```
names(ds)[names(ds)=="old1"] <- "new1"
names(ds)[names(ds)=="old2"] <- "new2"
```

or

```
ds <- within(ds, {new1 <- old1; new2 <- old2; rm(old1, old2)})
```

or

```
library(reshape)
ds <- rename(ds, c("old1"="new1", "old2"="new2"))
```

Note: The `names()` function provides a list of names associated with an object (see B.4.5). It is an efficient way to undertake this task, as it involves no copying of data (just a remapping of the names). The `edit()` function can be used to view names and edit values.

1.3.5 Add comment to a dataset or variable

HELP example: see 1.13.1

To help facilitate proper documentation of datasets, it can be useful to provide some annotation or description.

SAS

```
data ds (label="This is a comment about the dataset");
...
```

Note: The label can be viewed using `proc contents` (1.3.2) and retrieved as data using `ODS` (see A.7).

R

```
comment(ds) <- "This is a comment about the dataset"
```

Note: The `attributes()` function (see B.4.6) can be used to list all attributes, including any `comment()`, while the `comment()` function without an argument on the right hand side will display the comment, if present.

1.4 Derived variables and data manipulation

This section describes the creation of new variables as a function of existing variables in a dataset.

1.4.1 Create string variables from numeric variables

SAS

```
data ...;
   stringx = input(numericx, $char.);
run;
```

Note: Applying any string function to a numeric variable will force it to be treated as a character variable. As an example, concatenating (see 1.4.5) two numeric variables (i.e.

v3 = v1||v2) will result in a string . See A.6.4 for a discussion of informats, which apply variable types when reading in data.

R

```
stringx <- as.character(numericx)
typeof(stringx)
typeof(numericx)
```

Note: The typeof() function can be used to verify the type of an object; possible values include logical, integer, double, complex, character, raw, list, NULL, closure (function), special and builtin (see also section B.4.6).

1.4.2 Create numeric variables from string variables

SAS

```
data ...;
   numericx = input(stringx, integer.decimal);
run;
```

Note: In the argument to the input function, integer is the number of characters in the string, while decimal is an optional specification of how many characters appear after the decimal.

Applying any numeric function to a variable will force it to be treated as numeric. For example: a numericx = stringx * 1.0 statement will also make numericx a numeric variable.
See also A.6.4 for a discussion of informats, which apply variable types when reading in data.

R

```
numericx <- as.numeric(stringx)
typeof(stringx)
typeof(numericx)
```

Note: The typeof() function can be used to verify the type of an object (see 1.4.1 and B.4.6).

1.4.3 Extract characters from string variables

SAS

```
data ...;
   get2through4 = substr(x, 2, 3);
run;
```

Note: The syntax functions as follows: name of the variable, start character place, how many characters to extract. The last parameter is optional. When omitted, all characters after the location specified in the second space will be extracted.

R

```
get2through4 <- substr(x, 2, 4)
```

Note: The arguments to substr() specify the input vector, start character position and end character position.

1.4.4 Length of string variables

SAS

```
data ...;
   len = length(stringx);
run;
```

Note: In this example, `len` is a variable containing the number of characters in `stringx` for each observation in the dataset, excluding trailing blanks. Trailing blanks can be included through use of the `lengthc` function.

R

```
len <- nchar(stringx)
```

Note: The `nchar()` function returns a vector of lengths of each of the elements of the string vector given as argument, as opposed to the `length()` function (section 1.4.15) returns the number of elements in a vector.

1.4.5 Concatenate string variables

SAS

```
data ...;
   newcharvar = x1 || " VAR2 " x2;
run;
```

Note: The above SAS code creates a character variable `newcharvar` containing the character variable X_1 (which may be coerced from a numeric variable) followed by the string `" VAR2 "` then the character variable X_2. By default, no spaces are added.

R

```
newcharvar <- paste(x1, " VAR2 ", x2, sep="")
```

Note: The above R code creates a character variable `newcharvar` containing the character vector X_1 (which may be coerced from a numeric object) followed by the string `" VAR2 "` then the character vector X_2. The `sep=""` option leaves no additional separation character between these three strings.

1.4.6 Find strings within string variables

SAS

```
data ...;
   /* where is the first occurrence of "pat"? */
   match = find(stringx, "pat");
   /* where is the first occurrence of "pat" after startpos? */
   matchafter = find(stringx, "pat", startpos);
  /* how many times does "pat" appear? */
   howmany = count(stringx, "pat");
run;
```

Note: Without the option `startpos`, `find` returns the start character for the first appearance of `pat`. If `startpos` is positive, the search starts at `startpos`, if it is negative, the search is to the left, starting at `startpos`. If `pat` is not found or `startpos=0`, then `match=0`.

R

```
matches <- grep("pat", stringx)
positions <- regexpr("pat", stringx)
```

```
> x <- c("abc", "def", "abcdef", "defabc")
> grep("abc", x)
[1] 1 3 4
> regexpr("abc", x)
[1]  1 -1  1  4
attr(,"match.length")
[1]  3 -1  3  3
> regexpr("abc", x) < 0
[1] FALSE  TRUE FALSE FALSE
```

Note: The function `grep()` returns a list of elements in the vector given by `stringx` that match the given pattern, while the `regexpr()` function returns a numeric list of starting points in each string in the list (with -1 if there was no match). Testing `positions < 0` generates a vector of binary indicator of matches (TRUE=no match, FALSE=a match).

The regular expressions supported within `grep` and other related routines are quite powerful. For an example, Boolean `OR` expressions can be specified using the `|` operator. A comprehensive description of these can be found using `help(regex)`.

1.4.7 Remove spaces around string variables

SAS

```
data ...;
   noleadortrail = strip(stringx);
run;
```

Note: Leading blanks only can be removed with `left(stringx)`.

R

```
noleadortrail <- sub(' +$', '', sub('^ +', '', stringx))
```

Note: The arguments to `sub()` consist of a regular expression, a substitution value and a vector. In the first step, leading spaces are removed, then a separate call to `sub()` is used to remove trailing spaces (in both cases replacing the spaces with the null string. If instead of spaces all trailing whitespaces (e.g., tabs, space characters) should be removed, the regular expression `' +$'` should be replaced by `'[[:space:]]+$'`.

1.4.8 Upper to lower case

SAS

```
data ...;
   lowercasex = lowcase(x);
run;
```

or

```
data ...;
   lowercasex = translate(x, "abcdefghijklmnopqrstuvwxzy",
      "ABCDEFGHIJKLMNOPQRSTUVWXYZ") ;
run;
```

Note: The upcase function makes all characters upper case. Arbitrary translations from sets of characters can be made using the translate function.

R

```
lowercasex <- tolower(x)
```

or

```
lowercasex <- chartr("ABCDEFGHIJKLMNOPQRSTUVWXYZ",
       "abcdefghijklmnopqrstuvwxzy", x)
```

Note: The toupper() function can be used to convert to upper case. Arbitrary translations from sets of characters can be made using the chartr() function.

1.4.9 Create categorical variables from continuous variables

SAS *HELP example:* see 1.13.3 and 4.6.6

```
data ...;
   if x ne . then newcat = (x ge minval) + (x ge cutpoint1) +
                           ... + (x ge cutpointn);
run;
```

Note: Each expression within parentheses is a logical test returning 1 if the expression is true, 0 otherwise. If the initial condition is omitted then a missing value for x will return the value of 0 for newcat. More information about missing value coding can be found in section 1.4.14 (see 1.11.2 for more about conditional execution).

R

```
newcat1 <- (x >= minval) + (x >= cutpoint1) + ... + (x >= cutpointn)
```

Note: Each expression within parentheses is a logical test returning 1 if the expression is true, 0 if not true, and NA if x is missing. More information about missing value coding can be found in section 1.4.14.

1.4.10 Recode a categorical variable

A categorical variable may need to be recoded to have fewer levels.

SAS

```
data ...;
   newcat = (oldcat in (val1, val2, ..., valn)) +
            (oldcat in (...)) + ...;
run;
```

Note: The in function can also accept quoted strings as input. It returns a value of 1 if any of the listed values is equal to the tested value.

R

```
tmpcat <- oldcat
tmpcat[oldcat==val1] <- newval1
tmpcat[oldcat==val2] <- newval1
...
tmpcat[oldcat==valn] <- newvaln
newcat <- as.factor(tmpcat)
```

or

```
newcat <- cut(x, breaks=c(val2, ..., valn),
   labels=c("Cut1", "Cut2", ..., "Cutn"), right=FALSE)
```

Note: Creating the variable can be undertaken in multiple steps. A copy of the old variable is first made, then multiple assignments are made for each of the new levels, for observations matching the condition inside the index (see section B.4.2). In the final step, the categorical variable is coerced into a factor (class) variable. Alternatively, the `cut()` function can be used to create the factor vector in one operation, by specifying the cut-scores and the labels.

1.4.11 Create a categorical variable using logic

HELP example: see 1.13.3

Here we create a trichotomous variable `newvar` which takes on a missing value if the continuous nonnegative variable `oldvar` is less than 0, 0 if the continuous variable is 0, value 1 for subjects in group A with values greater than 0 but less than 50 and for subjects in group B with values greater than 0 but less than 60, or value 2 with values above those thresholds.

More information about missing value coding can be found in section 1.4.14.

SAS

```
data ...;
   if oldvar le 0 then newvar=.;
   else if oldvar eq 0 then newvar=0;
   else if (oldvar lt 50 and group eq "A") or
           (oldvar lt 60 and group eq "B")
      then newvar=1;
   else newvar=2;
run;
```

R

```
tmpvar <- rep(NA, length(oldvar))
tmpvar[oldvar==0] <- 0
tmpvar[oldvar>0 & oldvar<50 & group=="A"] <- 1
tmpvar[oldvar>0 & oldvar<60 & group=="B"] <- 1
tmpvar[oldvar>=50 & group=="A"] <- 2
tmpvar[oldvar>=60 & group=="B"] <- 2
newvar <- as.factor(tmpvar)
```

Note: Creating the variable is undertaken in multiple steps in this example. A vector of the correct length is first created containing missing values. Values are updated if they match the conditions inside the vector index (see section B.4.2). Care needs to be taken in the comparison of `oldvar==0` if non-integer values are present (see 1.8.5).

1.4.12 Formatting values of variables

HELP example: see 3.7.2

Sometimes it is useful to display category names that are more descriptive than variable names. In general, we do not recommend using this feature (except potentially for graphical output), as it tends to complicate communication between data analysts and other readers of output (see also labeling variables, 1.4.13). In this example, character labels are associated with a numeric variable (0=Control, 1=Low Dose and 2=High Dose).

SAS

```
proc format;
   value dosegroup 0 = 'Control' 1 = 'Low Dose' 2 = 'High Dose';
run;
```

Note: Many procedures accept a `format x dosegroup.` statement (note trailing '.'); this syntax will accept formats designed by the user with the `proc format` statement, as well as built-in formats (see 1.2.4). Categorizations of a variable can also be imposed using `proc format`, but this can be cumbersome. In all cases, a new variable should be created as described in 1.4.9 or 1.4.10.

R

```
x <- c(0, 0, 1, 1, 2)
x <- factor(x, 0:2, labels=c("Control", "Low Dose", "High Dose"))
```

Note: For this example, the command `x` returns:

```
  Control   Control  Low Dose  Low Dose High Dose
```

Additionally, the `names()` function can be used to associate a variable with the identifier (which is by default the observation number). As an example, this can be used to display the name of a region with the value taken by a particular variable measured in that region.

1.4.13 Label variables

As with the values of the categories, sometimes it is desirable to have a longer, more descriptive variable name (see also formatting variables, 1.4.12). In general, we do not recommend using this feature, as it tends to complicate communication between data analysts and other readers of output (a possible exception is in graphical output).

SAS

```
data ds;
   ...
   label x="This is the label for the variable 'x'";
run;
```

Note: The label is displayed instead of the variable name in all procedure output (except `proc print`, unless the `label` option is used) and can also be seen in `proc contents` (section 1.3.2).

Some procedures also allow `label` statements with identical syntax, in which case the label is used only for that procedure.

R

```
comment(x) <- "This is the label for the variable 'x'"
```

Note: The label for the variable can be extracted using `comment(x)` with no assignment or via `attribute(x)$comment`.

1.4.14 Account for missing values

HELP example: see 1.13.3

Missing values are ubiquitous in most real-world investigations. Both SAS and R feature support for missing value codes, though there are important distinctions which need to be kept in mind by an analyst, particularly when deriving new variables or fitting models.

In SAS, the default missing value code for numeric data is '.', which has a numeric value of negative infinity. There are 27 other pre-defined missing value codes (`._`, `.a` ... `.z`), which can be used, for example, to record different reasons for missingness. The missing value code for character data, for assignment, is `" "` (quote blank quote), displayed as a blank.

Listwise deletion is the default behavior for most multivariate procedures in SAS. That is, observations with missing values for any variables named in the procedure are omitted from all calculations. Data step functions are different: functions defined with mathematical operators (`+ - / * **`) will result in a missing value if any operand has a missing value, but named functions, such as `mean(x1, x2)` will result in the function as applied to the non-missing values.

In R, missing values are denoted by `NA`, a logical constant of length 1 which has no numeric equivalent. The missing value code is distinct from the character string value `"NA"`. The default behavior for most R functions is to return `NA` if any of the input vectors have any missing values.

SAS

```
data ds;
   missing = (x1 eq .);
   x2 = (1 + 2 + .)/3;
   x3 = mean(1, 2, .);

   if x4 = 999 then x4 = .;

   x5 = n(1, 2, 49. 123, .);
   x6 = nmiss(x2,x3);

   if x1 ne .;
   if x1 ne . then output;
run;
```

Note: The variable `missing` has a value of 1 if X_1 is missing, 0 otherwise. X_2 has a missing value, while $x_3 = 1.5$. Values of x_4 that were previously coded 999 are now marked as missing. The `n` function returns the number of non-missing values; X_5 has a value of 3; the `nmiss` function returns the number of missing values and here has a value of 1. The last two statements have identical meanings. They will remove all observations for which X_1 contains missing values.

R

```
> mean(c(1, 2, NA))
[1] NA
> mean(c(1, 2, NA), na.rm=TRUE)
[1] 1.5
> sum(na.omit(c(1, 2, NA)))
[1] 3
> x <- c(1, 3, NA)
> sum(!is.na(x))
[1] 2
> mean(x)
[1] NA
> mean(x, na.rm=TRUE)
[1] 2
```

The na.rm option is used to override the default behavior and omit missing values and calculate the result on the complete cases (this or related options are available for many functions). The ! (not) operator allows counting of the number of observed values (since is.na() returns a logical set to TRUE if an observation is missing). Values can be recoded to missing, as well as omitted (see B.3).

```
# remap values of x with missing value code of 999 to missing
x[x==999] <- NA
```

or

```
# set 999's to missing
is.na(x) <- x==999
# returns a vector of logicals
is.na(x)
# removes observations that are missing on that variable
na.omit(x)
# removes observations that are missing any variable
na.omit(ds)

library(Hmisc)
# display patterns of missing variables in a dataframe
na.pattern(ds)
```

Note: The default of returning NA for functions operating on objects with missing values can be overridden using options for a particular function by using na.omit(), adding the na.rm=TRUE option (e.g., for the mean() function) or specifying an na.action() (e.g., for the lm() function). Common na.action() functions include na.exclude(), na.omit(), and na.fail(). Arbitrary numeric missing values (999 in this example) can be mapped to R missing value codes using indexing and assignment. Here all values of x that are 999 are replaced by the missing value code of NA. The is.na() function returns a logical vector with TRUE corresponding to missing values (code NA in R). Input functions like scan() and read.table() have the default argument na.strings="NA". This can be used to recode on input for situations where a numeric missing value code has been used. R has other kinds of "missing" values, corresponding to floating point standards (see also is.infinite() and is.nan()).

The na.pattern() function can be used to determine the different patterns of missing values in a dataset. The na.omit() function returns the dataframe with missing values omitted (if a value is missing for a given row, all observations are removed, aka listwise deletion). More sophisticated approaches to handling missing data are generally recommended (see 6.5).

1.4.15 Observation number

SAS

HELP example: see 1.13.2

```
data ...;
   x = _n_;
run;
```

Note: The variable _n_ is created automatically by SAS and counts the number of lines of data that have been input into the data step. It is a temporary variable that it is not stored in the dataset unless a new variable is created (as demonstrated in the above code).

R

```
> y <- c("abc", "def", "ghi")
> x <- 1:length(y)
> x
[1] 1 2 3
```

Note: The `length()` function returns the number of elements in a vector. This can be used in conjunction with the : operator (section 1.11.3) to create a vector with the integers from 1 to the sample size. Observation numbers might also be set as case labels as opposed to the row number (see `names()`).

1.4.16 Unique values

HELP example: see 1.13.2

SAS

```
proc sort data=ds out=newds nodupkey;
   by x1 ... xk;
run;
```

Note: The dataset `newds` contains all the variables in the dataset `ds`, but only one row for each unique value across $x_1 x_2 \ldots x_k$.

R

```
uniquevalues <- unique(x)
uniquevalues <- unique(data.frame(x1, ..., xk))
```

Note: The `unique()` function returns each of the unique values represented by the vector or dataframe denoted by `x`.

1.4.17 Lagged variable

A lagged variable has the value of that variable in a previous row (typically the immediately previous one) within that dataset. The value of lag for the first observation will be missing (see 1.4.14).

SAS

```
data ...;
   xlag1 = lag(x);
run;
```

or

```
data ...;
   xlagk = lagk(x);
run;
```

Note: In the latter case, the variable `xlagk` contains the value of x from the kth preceding observation. The value of k can be any integer less than 101: the first `k` observations will have a missing value.

If executed conditionally, only observations with computed values are included. In other words, the statement `if (condition) then xlag1 = lag(x)` results in the variable `xlag1` containing the value of `x` from the most recently processed observation *for which* `condition` *was true*. This is a common cause of confusion.

R

```
lag1 <- c(NA, x[1:(length(x)-1)])
```

Note: This expression creates a one observation lag, with a missing value in the first position, and the first through second to last observation for the remaining entries. We can write a function to create lags of more than one observation.

```
lagk <- function(x, k) {
    len <- length(x)
    if (!floor(k)==k) {
        cat("k must be an integer")
    } else if (k<1 | k>(len-1)) {
        cat("k must be between 1 and length(x)-1")
    } else {
        return(c(rep(NA, k), x[1:(len-k)]))
    }
}
```

```
> lagk(1:10, 5)
 [1] NA NA NA NA NA  1  2  3  4  5
```

1.4.18 SQL

Structured Query Language (SQL) is a language for querying and modifying databases. SAS supports access to SQL through `proc sql`, while in R the `RMySQL`, `RSQLite` or `sqldf` packages can be used.

1.4.19 Perl interface

Perl is a high-level general purpose programming language [77]. SAS 9.2 supports Perl regular expressions in the data step via the `prxparse`, `prxmatch`, `prxchange`, `prxparen`, and `prxposn` functions. Details on their use can be found in the on-line help: Contents; SAS Products; Base SAS; SAS 9.2 Language Reference:Dictionary; Functions and CALL Routines under the names listed above. The `RSPerl` package provides a bidirectional interface between Perl and R.

1.5 Merging, combining, and subsetting datasets

A common task in data analysis involves the combination, collation, and subsetting of datasets. In this section, we review these techniques for a variety of situations.

1.5.1 Subsetting observations

SAS *HELP example:* see 1.13.4

```
data ...;
   if x eq 1;
run;
```

or

```
data ...;
   where x eq 1;
run;
```

or

```
data ...;
set ds (where= (x eq 1));
run;
```

Note: These examples create a new dataset consisting of observations where $x = 1$. The `if` statement has an implied "then output." The `where` syntax also works within procedures to limit the included observations to those that meet the condition, without creating a new dataset; see 4.6.9.

R

```
smallds <- ds[x==1,]
```

Note: This example creates a subject of a dataframe consisting of observations where $X = 1$. In addition, many functions allow specification of a `subset=expression` option to carry out a procedure on observations that match the expression (see 5.6.5).

1.5.2 Random sample of a dataset

See also random number seed (1.10.9)

It is sometimes useful to sample a subset (here quantified as *nsamp*) of observations without replacement from a larger dataset.

SAS

```
data ds2;
set ds;
   order = uniform(0);
run;

proc sort data=ds2;
   by order;
run;

data ds3;
set ds2;
   if _n_ le nsamp;
run;
```

Note: Note that if the second `data` step is omitted, the observations have been randomly reordered.

It is also possible to generate a random sample in a single data step by generating a uniform random variate for each observation in the original data but using an `if` statement to retain only those which meet a criteria which changes with the number retained.

R

```
# permutation of a variable
newx <- sample(x, replace=FALSE)

# permutation of a dataset
obs <- sample(1:dim(ds)[1], dim(ds)[1], replace=FALSE)
newds <- ds[obs,]
```

Note: By default, the `sample()` function in R takes a sample of all values (determined in this case by determining the number of observations in `ds`), without replacement. This is equivalent to a permutation of the order of values in the vector. The `replace=TRUE` option can be used to override this (e.g., when bootstrapping, see section 2.1.8). Fewer values can be sampled by specifying the `size` option.

1.5.3 Convert from wide to long (tall) format

HELP example: see 4.6.9

Sometimes data are available in a different shape than that required for analysis. One example of this is commonly found in repeated longitudinal measures studies. In this setting it is convenient to store the data in a wide or multivariate format with one line per subject, containing typically subject invariant factors (e.g., gender), as well as a column for each repeated outcome. An example would be:

```
id female inc80 inc81 inc82
1    0    5000  5500  6000
2    1    2000  2200  3300
3    0    3000  2000  1000
```

where the income for 1980, 1981, and 1982 are included in one row for each id.

In contrast, SAS and R tools for repeated measures analyses (4.2.2) typically require a row for each repeated outcome, such as

```
id year female inc
1  80   0      5000
1  81   0      5500
1  82   0      6000
2  80   1      2000
2  81   1      2200
2  82   1      3300
3  80   0      3000
3  81   0      2000
3  82   0      1000
```

In this section and in section (1.5.4) below, we show how to convert between these two forms of this example data.

SAS

```
data long;
set wide;
   array incarray [3] inc80 - inc82;
   do year = 80 to 82;
      inc = incarray[year - 79];
      output;
   end;
   drop inc80 - inc82;
run;
```

or

```
data long;
set wide;
   year=80; inc=inc80; output;
   year=81; inc=inc81; output;
   year=82; inc=inc82; output;
   drop inc80 - inc82;
run;
```

or

```
proc transpose data=wide out=long_i;
   var inc80 - inc82;
   by id female;
run;

data long;
set long_i;
   year=substr(_name_, 4, 2)*1.0;
   drop _name_;
   rename col1=inc;
run;
```

Note: The `year=substr()` statement in the last data step is required if the value of `year` must be numeric. The remainder of that step makes the desired variable name appear, and removes extraneous information.

R

```
long <- reshape(wide, idvar="id", varying=list(names(wide)[3:5]),
   v.names="inc", timevar="year", times=80:82, direction="long")
```

Note: The list of variables to transpose is provided in the list `varying`, creating `year` as the time variable with values specified by `times` (see also `library(reshape)` for more flexible dataset transformations).

1.5.4 Convert from long (tall) to wide format

HELP example: see 4.6.9

See also section 1.5.3 (reshape from wide to tall)

SAS

```
proc transpose data=long out=wide (drop=_name_) prefix=inc;
   var inc;
   id year;
   by id female;
run;
```

Note: The `(drop=_name_)` option prevents the creation of an unneeded variable in the `wide` dataset.

R

```
wide <- reshape(long, v.names="inc", idvar="id", timevar="year",
   direction="wide")
```

Note: This example assumes that the dataset `long` has repeated measures on `inc` for subject `id` determined by the variable `year`. See also `library(reshape)` for more flexible dataset transformations.

1.5.5 Concatenate datasets

SAS

```
data newds;
set ds1 ds2;
run;
```

Note: The datasets `ds1` and `ds2` are assumed to previously exist. The newly created dataset `newds` has as many rows as the sum of rows in `ds1` and `ds2`, and as many columns as unique variable names across the two input datasets.

R

```
newds <- rbind(ds1, ds2)
```

Note: The result of `rbind()` is a dataframe with as many rows as the sum of rows in `ds1` and `ds2`. Data frames given as arguments to `rbind()` must have the same column names. The similar `cbind()` function makes a dataframe with as many columns as the sum of the columns in the input objects.

1.5.6 Sort datasets

HELP example: see 1.13.4

SAS

```
proc sort data=ds;
   by x1 ... xk;
run;
```

Note: The keyword `descending` can be inserted before any variable to sort that variable from high to low (see also A.6.2).

R

```
sortds <- ds[order(x1, x2, ..., xk),]
```

Note: The R command `sort()` can be used to sort a vector, while `order()` can be used to sort dataframes by selecting a new permutation of order for the rows. The `decreasing` option can be used to change the default sort order (for all variables). The command `sort(x)` is equivalent to `x[order(x)]`. As an alternative, a numeric variable can be reversed by specifying `-x1` instead of `x1`.

1.5.7 Merge datasets

HELP example: see 4.6.11

Merging datasets is commonly required when data on single units are stored in multiple tables or datasets. We consider a simple example where variables `id, year, female` and `inc` are available in one dataset, and variables `id` and `maxval` in a second. For this simple example, with the first dataset given as:

```
id year female inc
1  80   0      5000
1  81   0      5500
1  82   0      6000
2  80   1      2000
2  81   1      2200
2  82   1      3300
3  80   0      3000
3  81   0      2000
3  82   0      1000
```

and the second given below.

```
id maxval
2  2400
1  1800
4  1900
```

The desired merged dataset would look like:

```
   id year female  inc maxval
1   1   81      0 5500   1800
2   1   80      0 5000   1800
3   1   82      0 6000   1800
4   2   82      1 3300   2400
5   2   80      1 2000   2400
6   2   81      1 2200   2400
7   3   82      0 1000     NA
8   3   80      0 3000     NA
9   3   81      0 2000     NA
10  4   NA     NA   NA   1900
```

in R, or equivalently, as below in SAS.

```
   id year female  inc maxval
1   1   81      0 5500   1800
2   1   80      0 5000   1800
3   1   82      0 6000   1800
4   2   82      1 3300   2400
5   2   80      1 2000   2400
6   2   81      1 2200   2400
7   3   82      0 1000      .
8   3   80      0 3000      .
9   3   81      0 2000      .
10  4    .      .    .   1900
```

SAS

```
proc sort data=ds1; by x1 ... xk;
run;

proc sort data=ds2; by x1 ... xk;
run;

data newds;
merge ds1 ds2;
   by x1 ... xk;
run;
```

For example, the result desired in the note above can be created as follows, assuming the two datasets are named `ds1` and `ds2`:

```
proc sort data=ds1; by id; run;

proc sort data=ds2; by id; run;

data newds;
merge ds1 ds2;
   by id;
run;
```

Note: The by statement in the data step describes the matching criteria, in that every observation with a unique set of X_1 through X_k in ds1 will be matched to every observation with the same set of X_1 through X_k in ds2. The output dataset will have as many columns as there are uniquely named variables in the input datasets, and as many rows as unique values across X_1 through X_k. The by statement can be omitted, which results in the nth row of each dataset contributing to the nth row of the output dataset, though this is rarely desirable. If matched rows have discrepant values for a commonly-named variable, the value in the later-named dataset is used in the output dataset.

R

```
newds <- merge(ds1, ds2, by=id, all=TRUE)
```

Note: The all option specifies that extra rows will be added to the output for any rows that have no matches in the other dataset. Multiple variables can be specified in the by option; if this is left out all variables in both datasets are used: see help(merge).

1.5.8 Drop variables in a dataset

HELP example: see 1.13.1

It is often desirable to prune extraneous variables from a dataset to simplify analyses.

SAS

```
data ds;
   ...
   keep x1 xk;
   ...
run;
```

or

```
data ds;
set old_ds (keep=x1 xk);
...
run;
```

Note: The complementary syntax drop can also be used, both as a statement in the data step and as a data statement option.

R

```
ds[,c("x1", "xk")]
```

Note: The above example created a new dataframe consisting of the variables x1 and xk. An alternative is to specify the variables to be excluded (in this case the second):

```
ds[,names(ds)[-2]]

or

ds[,-2]
```

More sophisticated ways of listing the variables to be kept are available. For example, the command `ds[,grep("x1|^pat", names(ds))]` would keep `x1` and all variables starting with `pat` (see also 1.4.6).

1.6 Date and time variables

For SAS, variables in the date formats are integers counting the number of days since January 1, 1960. In R, the date functions return a `Date` class that represents the number of days since January 1, 1970. The R function `as.numeric()` can be used to create a numeric variable with the number of days since 1/1/1970 (see also the `Chron` package).

1.6.1 Create date variable

SAS

```
data ...;
   dayvar = input("04/29/2010", mmddyy10.);
   todays_date = today();
run;
```

Note: `dayvar` is the integer number of days between January 1, 1960 and April 29, 2010. The value of `todays_date` is the integer number of days between January 1, 1960 and the day the current instance of SAS was opened.

R

```
dayvar <- as.Date("2010-04-29")
todays_date <- as.Date(Sys.time())
```

Note: The return value of `as.Date()` is a `Date` class object. If converted to numeric `dayvar`, it represents the number of days between January 1, 1970 and April 29, 2010, while `todays_date` is the integer number of days since January 1, 1970.

1.6.2 Extract weekday

SAS

```
data ...;
   wkday = weekday(datevar);
run;
```

Note: The `weekday` function returns an integer representing the weekday, 1=Sunday, ..., 7=Saturday.

R

```
wkday <- weekdays(datevar)
```

Note: `wkday` contains a string with name of the weekday of the `Date` object.

1.6.3 Extract month

SAS

```
data ...;
   monthval = month(datevar);
run;
```

Note: The `month` function returns an integer representing the month, 1=January, ..., 12=December.

R

```
monthval <- months(datevar)
```

Note: The function `months()` returns a string with the name of the month of the `Date` object.

1.6.4 Extract year

SAS

```
data ...;
   yearval = year(datevar);
run;
```

Note: The variable `yearval` is years counted in the Common Era (CE, also called AD).

R

```
yearval <- substr(as.POSIXct(datevar), 1, 4)
```

Note: The `as.POSIXct()` function returns a string representing the date, with the first four characters corresponding to the year.

1.6.5 Extract quarter

SAS

```
data ...;
   qtrval = qrt(datevar);
run;
```

Note: The return values for `qtrval` are 1, 2, 3, or 4.

R

```
qrtval <- quarters(datevar)
```

Note: The function `quarters()` returns a string representing the quarter of the year (e.g., `"Q1"` or `"Q2"`) given by the `Date` object.

1.6.6 Create time variable

See also 1.7.1 (timing commands)

SAS

```
data ...;
   timevar_1960 = datetime();
   timevar_midnight = time();
run;
```

Note: The variable `timevar_1960` contains the number of seconds since midnight, December 31, 1959. The variable `timevar_midnight` contains the number of seconds since the most recent midnight.

R

```
> arbtime <- as.POSIXlt("2010-04-29 17:15:45 NZDT")
> arbtime
[1] "2010-04-29 17:15:45"
> Sys.time()
[1] "2010-04-01 10:12:11 EST"
```

Note: The objects `arbtime` and `now` can be compared with the subtraction operator to monitor elapsed time.

1.7 Interactions with the operating system

1.7.1 Timing commands

SAS

```
options stimer;
options fullstimer;
```

Note: These options request that a subset (`stimer`) or all available (`fullstimer`) statistics are reported in the SAS log. We are not aware of a simple way to get the statistics except by reading the log.

R

```
system.time(expression)
```

Note: The `expression` (e.g., call to any user or system defined function, see B.4.1) given as argument to the `system.time()` function is evaluated, and the user, system, and total (elapsed) time is returned.

1.7.2 Execute command in operating system

SAS

```
x;
```

or

```
x 'OS command';
```

or

```
data ...;
   call system("OS command");
run;
```

Note: An example command statement would be `x 'dir'`. The statement consisting of just `x` will open a command window. Related statements are `x1`, `x2`, ... , `x9`, which allow up to 9 separate operating system tasks to be executed simultaneously.

The `x` command need not be in a data step, and cannot be executed conditionally. In other words, if it appears as a consequence in an `if` statement, it will be executed regardless of whether or not the test in the `if` statement is true or not. Use the `call system` statement as shown to execute conditionally. This syntax to open a command window may not be available in all operating systems.

R

```
system("ls")
```

Note: The command `ls` lists the files in the current working directory (see 1.7.5 to capture this information). When R is running under Windows, the `shell()` command can be used to start a command window.

1.7.3 Find working directory

SAS

```
x;
```

Note: This will open a command window; the current directory in this window is the working directory. The working directory can also be found using the method shown in section (1.7.5) using the `cd` command in Windows or the `pwd` command in Linux.

The current directory is displayed by default in the status line at the bottom of the SAS window.

R

```
getwd()
```

Note: The command `getwd()` displays the current working directory.

1.7.4 Change working directory

SAS

```
x 'cd dir_location';
```

Note: This can also be done interactively by double-clicking the display of the current directory in the status line at the bottom of the SAS window (note that this applies for Windows installations, for other operating systems, see the on-line help: Contents; Using SAS software in Your Operating Environment; SAS 9.2 companion for <your OS>; Running SAS under <your OS>).

R

```
setwd("dir_location")
```

Note: The command `setwd()` changes the current working directory to the (absolute or relative) pathname given as argument. This can also be done interactively under Windows and Mac OS X by selecting the `Change Working Directory` option under the `Misc` menu.

1.7.5 List and access files

SAS

```
filename filehandle pipe 'dir /b'; /* Windows */
filename filehandle pipe 'ls'; /* Unix or Mac OS X */

data ds;
   infile filehandle truncover;
   input x $20.;
run;
```

Note: The `pipe` is a special file type which passes the characters inside the single quote to the operating system when read using the `infile` statement, then reads the result. The above code lists the contents of the current directory. The dataset `ds` contains a single character variable `x` with the file names. The file handle can be no longer than 8 characters.

R

```
list.files()
```

Note: The `list.files()` command returns a character vector of filenames in the current directory (by default). Recursive listings are also supported. The function `file.choose()` provides an interactive file browser, and can be given as an argument to functions such as `read.table()` (section 1.1.2) or `read.csv()` (section 1.1.4). Related file operation functions include `file.access()`, `file.info()` and `files()`.

1.8 Mathematical functions

1.8.1 Basic functions

SAS

```
data ...;
   minx = min(x1, ..., xk);
   maxx = max(x1, ..., xk);
   meanx = mean(x1, ..., xk);
   stddevx = std(x1, ..., xk);
   sumx = sum(x1, ..., xk)
   absolutevaluex = abs(x);
   etothex = exp(x);
   xtothey = x**y;
   squareroottx = sqrt(x);
   naturallogx = log(x);
   logbase10x = log10(x);
   logbase2x = log2(x);
run;
```

Note: The first five functions operate on a row-by-row basis within SAS (the equivalent within R operates on a column-wise basis).

R

```
minx <- min(x)
maxx <- max(x)
meanx <- mean(x)
stddevx <- sd(x)
absolutevaluex <- abs(x)
squarerootx <- sqrt(x)
etothex <- exp(x)
xtothey <- x^y
naturallogx <- log(x)
logbase10x <- log10(x)
logbase2x <- log2(x)
logbasearbx = log(x, base=42)
```

Note: The first five functions operate on a column-wise basis in R (the equivalent within SAS operates on a row-wise basis).

1.8.2 Trigonometric functions

SAS

```
data ...;
   sinx = sin(x);
   sinpi = sin(constant('PI'));
   cosx = cos(x);
   tanx = tan(x);
   arccosx = arcos(x);
   arcsinx = arsin(x);
   arctanx  = atan(x);
   arctanxy = atan2(x, y);
run;
```

R

```
sin(pi)
cos(0)
tan(pi/4)
acos(x)
asin(x)
atan(x)
atan2(x, y)
```

1.8.3 Special functions

SAS

```
data ...;
   betaxy = beta(x, y);
   gammax = gamma(x);
   factorialn = fact(n);
   nchooser = comb(n, r);
   npermr = perm(n, r);
run;
```

R

```
betaxy <- beta(x, y)
gammax <- gamma(x)
factorialn <- factorial(n)
nchooser <- choose(n, r)

library(gtools)
nchooser <- length(combinations(n, r)[,1])
npermr <- length(permutations(n, r)[,1])
```

Note: The `combinations()` and `permutations()` functions return a list of possible combinations and permutations: the count equivalent to the SAS functions above can be calculated through use of the `length()` function given the first column of the output.

1.8.4 Integer functions

See also 1.2.5 (rounding and number of digits to display)

SAS

```
data ...;
   nextintx = ceil(x);
   justintx = floor(x);
   roundx = round(x1, x2);
   roundint = round(x, 1);
   movetozero = int(x);
run;
```

Note: The value of `roundx` is X_1, rounded to the nearest X_2. The value of `movetozero` is the same as `justint` if $x > 0$ or `nextint` if $x < 0$.

R

```
nextintx <- ceiling(x)
justintx <- floor(x)
round2dec <- round(x, 2)
roundint <- round(x)
keep4sig <- signif(x, 4)
movetozero <- trunc(x)
```

Note: The second parameter of the `round()` function determines how many decimal places to round. The value of `movetozero` is the same as `justint` if $x > 0$ or `nextint` if $x < 0$.

1.8.5 Comparisons of floating point variables

Because certain floating point values of variables do not have exact decimal equivalents, there may be some error in how they are represented on a computer. For example, if the true value of a particular variable is 1/7, the approximate decimal is 0.1428571428571428. For some operations (for example, tests of equality), this approximation can be problematic.

SAS

```
data ds;
   x1 = ((1/7) eq .142857142857);
   x2 = (fuzz((1/7) - .142857142857) eq 0);
run;
```

Note: In the above example, $x_1 = 0$, $x_2 = 1$. If the argument to fuzz is less than 10^{-12} then the result is the nearest integer.

R

```
> all.equal(0.1428571, 1/7)
[1] "Mean relative difference: 3.000000900364093e-07"
> all.equal(0.1428571, 1/7, tolerance=0.0001)
[1] TRUE
```

Note: The tolerance option for the all.equal() function determines how many decimal places to use in the comparison of the vectors or scalars (the default tolerance is set to the underlying lower level machine precision).

1.8.6 Derivative

Rudimentary support for finding derivatives is available within R. These functions are particularly useful for high-dimensional optimization problems (see 1.8.7).

R

```
D(expression(x^3), "x")
```

Note: Second (or higher order) derivatives can be found by repeatedly applying the D function with respect to X. This function (as well as deriv()) are useful in numerical optimization (see the nlm(), optim() and optimize() functions).

1.8.7 Optimization problems

SAS and R can be used to solve optimization (maximization) problems. As an extremely simple example, consider maximizing the area of a rectangle with perimeter equal to 20. Each of the sides can be represented by x and 10-x, with area of the rectangle equal to $x * (10 - x)$.

SAS

```
proc iml;
   start f_area(x);
   f = x*(10-x);
   return (f);
   finish f_area;
   con = {0, 10};
   x = {2} ;
   optn = {1, 2};
   call nlpcg(rc, xres, "f_area", x, optn, con);
quit;
```

Note: The above uses conjugate gradient optimization. Several additional optimization routines are provided in proc iml (see the on-line help: Contents; SAS Products; SAS/IML User's Guide; Nonlinear Optimization Examples).

R

```
f <- function(x) { return(x*(10-x)) }
optimize(f, interval=c(0, 10), maximum=TRUE)
```

Note: Other optimization functions available within R include nlm(), uniroot(), optim() and constrOptim() (see also the CRAN Optimization and Mathematical Programming Task View).

1.9 Matrix operations

Matrix operations are often needed in statistical analysis. For SAS, `proc iml` (a separate product from SAS/STAT), is needed to treat data as a matrix. Within R, matrices can be created using the `matrix()` function (see B.4.4): matrix operations are then immediately available.

Here, we briefly outline the process needed to read a SAS dataset into SAS/IML as a matrix, perform some function, then make the result available as a SAS native dataset. Throughout this section, we use capital letters to emphasize that a matrix is described, though `proc iml` is not case-sensitive (unlike R).

```
proc iml;
   use ds;
   read all var(x1 ... xk) into Matrix_x;
   ...  /* perform a function of some sort */
   print Matrix_x;    /* print the matrix to the output window */
   create newds from Matrix_x;
   append from Matrix_x;
quit;
```

Note: Calls to `proc iml` end with a `quit` statement, rather than a `run` statement.

In addition to the routines described below, the `Matrix` library in R is particularly useful for manipulation of large as well as sparse matrices.

1.9.1 Create matrix

In this entry, we demonstrate creating a 2×2 matrix consisting of the first four nonzero integers:

$$A = \begin{pmatrix} 1 & 2 \\ 3 & 4 \end{pmatrix}.$$

SAS

```
proc iml;
   A = {1 2, 3 4};
quit;
```

R

```
A <- matrix(c(1, 2, 3, 4), 2, 2, byrow=TRUE)
```

1.9.2 Transpose matrix

SAS

```
proc iml;
   A = {1 2, 3 4};
   transA = A`;
   transA_2 = t(A);
quit;
```

Note: Both `transA` and `transA_2` contain the transpose of `A`.

R

```
A <- matrix(c(1, 2, 3, 4), 2, 2, byrow=TRUE)
transA <- t(A)
```

1.9.3 Invert matrix

SAS

```
proc iml;
   A = {1 2, 3 4};
   Ainv = inv(A);
quit;
```

R

```
A <- matrix(c(1, 2, 3, 4), 2, 2, byrow=TRUE)
Ainv <- solve(A)
```

1.9.4 Create submatrix

SAS

```
proc iml;
   A = {1 2 3 4, 5 6 7 8, 9 10 11 12};
   Asub = a[2:3, 3:4];
quit;
```

R

```
A <- matrix(1:12, 3, 4, byrow=TRUE)
Asub <- A[2:3, 3:4]
```

1.9.5 Create a diagonal matrix

SAS

```
proc iml;
   A = {1 2, 3 4};
   diagMat = diag(A);
quit;
```

Note: For matrix A, this results in a matrix with the same diagonals, but with all off-diagonals set to 0. For vector argument, the function generates a matrix with the vector values as the diagonals and all off-diagonals 0.

R

```
A <- matrix(c(1, 2, 3, 4), 2, 2, byrow=TRUE)
diagMat <- diag(c(1, 4))    # argument is a vector
diagMat <- diag(diag(A))    # A is a matrix
```

Note: For vector argument, the `diag()` function generates a matrix with the vector values as the diagonals and all off-diagonals 0. For matrix A, the `diag()` function creates a vector of the diagonal elements (see 1.9.6); a diagonal matrix with these diagonal entries, but all off-diagonals set to 0 can be created by running the `diag()` with this vector as argument.

1.9.6 Create vector of diagonal elements

SAS

```
proc iml;
   A = {1 2, 3 4};
   diagVals = vecdiag(A);
quit;
```

Note: The vector `diagVals` contains the diagonal elements of matrix `A`.

R

```
A <- matrix(c(1, 2, 3, 4), 2, 2, byrow=TRUE)
diagVals <- diag(A)
```

1.9.7 Create vector from a matrix

SAS

```
proc iml;
   A = {1 2, 3 4};
   newvec = shape(A, 1);
quit;
```

Note: This makes a row vector from all the values in the matrix.

R

```
A <- matrix(c(1, 2, 3, 4), 2, 2, byrow=TRUE)
newvec <- c(A)
```

1.9.8 Calculate determinant

SAS

```
proc iml;
   A = {1 2, 3 4};
   detval = det(A);
quit;
```

R

```
A <- matrix(c(1, 2, 3, 4), 2, 2, byrow=TRUE)
detval <- det(A)
```

1.9.9 Find eigenvalues and eigenvectors

SAS

```
proc iml;
   A = {1 2, 3 4};
   Aeval = eigval(A);
   Aevec = eigvec(A);
quit;
```

R

```
A <- matrix(c(1, 2, 3, 4), 2, 2, byrow=TRUE)
Aev <- eigen(A)
Aeval <- Aev$values
Aevec <- Aev$vectors
```

Note: The `eigen()` function in R returns a list consisting of the eigenvalues and eigenvectors, respectively, of the matrix given as argument.

1.9.10 Calculate singular value decomposition

The singular value decomposition of a matrix A is given by $A = U * \text{diag}(Q) * V^T$ where $U^T U = V^T V = V V^T = I$ and Q contains the singular values of A.

SAS

```
proc iml;
   A = {1 2, 3 4};
   call svd(U, Q, V, A);
quit
```

R

```
A <- matrix(c(1, 2, 3, 4), 2, 2, byrow=TRUE)
svdres <- svd(A)
U <- svdres$u
Q <- svdres$d
V <- svdres$v
```

Note: The `svd()` function returns a list with components corresponding to a vector of singular values, a matrix with columns corresponding to the left singular values, and a matrix with columns containing the right singular values.

1.10 Probability distributions and random number generation

SAS and R can calculate quantiles and cumulative distribution values as well as generate random numbers for a large number of distributions. Random variables are commonly needed for simulation and analysis. SAS includes comprehensive random number generation through the `rand` function, while R provides a series of `r`-commands.

Both packages allow specification of a seed for the random number generator. This is important to allow replication of results (e.g., while testing and debugging). Information about random number seeds can be found in section 1.10.9.

Table 1.1 summarizes support for quantiles, cumulative distribution functions, and random numbers. More information on probability distributions within R can be found in the CRAN Probability Distributions Task View.

1.10.1 Probability density function

Both R and SAS use similar syntax for a variety of distributions. Here we use the Normal distribution as an example; others are shown in Table 1.1 (page 43).

SAS

```
data ...;
   y = cdf('NORMAL', 1.96, 0, 1);
run;
```

R

```
y <- pnorm(1.96, 0, 1)
```

1.10.2 Quantiles of a probability density function

Both R and SAS use similar syntax for a variety of distributions. Here we use the Normal distribution as an example; others are shown in Table 1.1 (p. 43).

SAS

```
data ...;
   y = quantile('NORMAL', .975, 0, 1);
run;
```

R

```
y <- qnorm(.975, 0, 1)
```

1.10.3 Uniform random variables

SAS

```
data ...;
   x1 = uniform(seed);
   x2 = rand('UNIFORM');
run;
```

Note: The variables x_1 and x_2 are uniform on the interval (0,1). The `ranuni()` function is a synonym for `uniform()`.

R

```
x <- runif(n, 0, 1)
```

Note: The arguments specify the number of variables to be created and the range over which they are distributed.

1.10.4 Multinomial random variables

SAS

```
data ...;
   x1 = rantbl(seed, p1, p2, ..., pk);
   x2 = rand('TABLE', p1, p2, ..., pk);
run;
```

Note: The variables x_1 and x_2 take the value i with probability p_i and value $k+1$ with value $1 - \sum_{i=1}^{k} p_i$.

Table 1.1: Quantiles, probabilities, and pseudo-random number generation: distributions available in SAS and R

Distribution	R DISTNAME	SAS DISTNAME
Beta	`beta`	BETA
Beta-binomial	`betabin`*	
binomial	`binom`	BINOMIAL
Cauchy	`cauchy`	CAUCHY
chi-square	`chisq`	CHISQUARE
exponential	`exp`	EXPONENTIAL
F	`f`	F
gamma	`gamma`	GAMMA
geometric	`geom`	GEOMETRIC
hypergeometric	`hyper`	HYPERGEOMETRIC
inverse Normal	`inv.gaussian`*	IGAUSS+
Laplace	`laplace`*	LAPLACE
logistic	`logis`	LOGISTIC
lognormal	`lnorm`	LOGNORMAL
negative binomial	`nbinom`	NEGBINOMIAL
normal	`norm`	NORMAL
Poisson	`pois`	POISSON
Student's t	`t`	T
Uniform	`unif`	UNIFORM
Weibull	`weibull`	WEIBULL

Note: For R, prepend `d` to the command to compute quantiles of a distribution `dDISTNAME(xvalue, parm1, ..., parmn)`, `p` for the cumulative distribution function, `pDISTNAME(xvalue, parm1, ..., parmn)`, `q` for the quantile function `qDISTNAME(prob, parm1, ..., parmn)`, and `r` to generate random variables `rDISTNAME(nrand, parm1, ..., parmn)` where in the last case a vector of `nrand` values is the result. For SAS, random variates can be generated from the `rand` function: `rand('DISTNAME', parm1, ..., parmn)`, the areas to the left of a value via the `cdf` function: `cdf('DISTNAME', quantile, parm1, ..., parmn)`, and the quantile associated with a probability (the inverse CDF) via the `quantile` function: `quantile('DISTNAME', probability, parm1, ..., parmn)`, where the number of `parms` varies by distribution. Details are available through the on-line help: Contents; SAS Products; Base SAS; SAS 9.2 Language Reference: Dictionary; Dictionary of Language Elements; Functions and CALL Routines; RAND Function. Note that in this instance SAS is case-sensitive.

* The `betabin()`, `inv.gaussian()`, and `laplace()` families of distributions are available using `library(VGAM)`.

+ The inverse normal is not available in the `rand` function; inverse Normal variates can be generated by taking the inverse of Normal random variates.

R

```
library(Hmisc)
x <- rMultinom(matrix(c(p1, p2, ..., pr), 1, r), n)
```

Note: The function `rMultinom()` from the `Hmisc` library allows the specification of the desired multinomial probabilities ($\sum_r p_r = 1$) as a $1 \times r$ matrix. The final parameter is the number of variates to be generated See also `rmultinom()` in the `stats` package).

1.10.5 Normal random variables

HELP example: see 1.13.5

SAS

```
data ...;
   x1 = normal(seed);
   x2 = rand('NORMAL', mu, sigma);
run;
```

Note: The variable X_1 is a standard Normal ($\mu = 0$ and $\sigma = 1$), while X_2 is Normal with specified mean and standard deviation. The function `rannor()` is a synonym for `normal()`.

R

```
x1 <- rnorm(n)
x2 <- rnorm(n, mu, sigma)
```

Note: The arguments specify the number of variables to be created and (optionally) the mean and standard deviation (default $\mu = 0$ and $\sigma = 1$).

1.10.6 Multivariate normal random variables

For the following, we first create a 3×3 covariance matrix. Then we generate 1000 realizations of a multivariate Normal vector with the appropriate correlation or covariance.

SAS

```
data Sigma (type=cov);
infile cards;
input _type_ $ _Name_ $ x1 x2 x3;
cards;
cov    x1      3 1 2
cov    x2      1 4 0
cov    x3      2 0 5
;
run;

proc simnormal data=sigma out=outtest2 numreal=1000;
   var x1 x2 x3;
run;
```

Note: The `type=cov` option to the `data` step defines `Sigma` as a special type of SAS dataset which contains a covariance matrix in the format shown. A similar `type=corr` dataset can be used to generate using a correlation matrix instead of a covariance matrix.

R

```
library(MASS)
mu <- rep(0, 3)
Sigma <- matrix(c(3, 1, 2,
                  1, 4, 0,
                  2, 0, 5), nrow=3)
xvals <- mvrnorm(1000, mu, Sigma)
apply(xvals, 2, mean)
```

or

```
rmultnorm <- function(n, mu, vmat, tol=1e-07)
# a function to generate random multivariate Gaussians
{
   p <- ncol(vmat)
   if (length(mu)!=p)
      stop("mu vector is the wrong length")
   if (max(abs(vmat - t(vmat))) > tol)
      stop("vmat not symmetric")
   vs <- svd(vmat)
   vsqrt <- t(vs$v %*% (t(vs$u) * sqrt(vs$d)))
   ans <- matrix(rnorm(n * p), nrow=n) %*% vsqrt
   ans <- sweep(ans, 2, mu, "+")
   dimnames(ans) <- list(NULL, dimnames(vmat)[[2]])
   return(ans)
}
xvals <- rmultnorm(1000, mu, Sigma)
apply(xvals, 2, mean)
```

Note: The returned object `xvals`, of dimension 1000×3, is generated from the variance covariance matrix denoted by `Sigma`, which has first row and column (3,1,2). An arbitrary mean vector can be specified using the `c()` function.

Several techniques are illustrated in the definition of the `rmultnorm` function. The first lines test for the appropriate arguments, and return an error if the conditions are not satisfied. The singular value decomposition (see 1.9.10) is carried out on the variance covariance matrix, and the `sweep` function is used to transform the univariate normal random variables generated by `rnorm` to the desired mean and covariance. The `dimnames()` function applies the existing names (if any) for the variables in `vmat`, and the result is returned.

1.10.7 Exponential random variables

SAS

```
data ...;
   x1 = ranexp(seed);
   x2 = rand('EXPONENTIAL');
run;
```

Note: The expected value of both X_1 and X_2 is 1: for exponentials with expected value k, multiply the generated value by k.

R

```
x <- rexp(n, lambda)
```

Note: The arguments specify the number of variables to be created and (optionally) the inverse of the mean (default $\lambda = 1$).

1.10.8 Other random variables

HELP example: see 1.13.5

The list of probability distributions supported within SAS and R can be found in Table 1.1, page 43. In addition to these distributions, the inverse probability integral transform can be used to generate arbitrary random variables with invertible cumulative density function F (exploiting the fact that $F^{-1} \sim U(0,1)$). As an example, consider the generation of random variates from an exponential distribution with rate parameter λ, where $F(X) = 1 - \exp(-\lambda X) = U$. Solving for X yields $X = -\log(1-U)/\lambda$. If we generate a Uniform(0,1) variable, we can use this relationship to generate an exponential with the desired rate parameter.

SAS

```
data ds;
   lambda = 2;
   uvar = uniform(42);
   expvar = -1 * log(1-uvar)/lambda;
run;
```

R

```
lambda <- 2
expvar <- -log(1-runif(1))/lambda
```

1.10.9 Setting the random number seed

SAS includes comprehensive random number generation through the `rand` function. For variables created this way, an initial seed is selected automatically by SAS based on the system clock. Sequential calls use a seed derived from this initial seed. To generate a replicable series of random variables, use the `call streaminit` function before the first call to `rand`.

SAS

```
call streaminit(42);
```

Note: A set of separate SAS functions for random number generation includes `normal`, `ranbin`, `rancau`, `ranexp`, `rangam`, `rannor`, `ranpoi`, `rantbl`, `rantri`, `ranuni`, and `uniform`. For these functions, calling with an argument of (0) is equivalent to calling the `rand` function without first running `call streaminit`; an initial seed is generated from the system clock. Calling the same functions with an integer greater than 0 as argument is equivalent to running `call streaminit` before an initial use of `rand`. In other words, this will result in a series of variates based on the first specified integer. Note that `call streaminit` or specifying an integer to one of the specific functions need only be performed once per `data` step; all seeds within that `data` step will be based on that seed.

In R, the default behavior is a seed based on the system clock. To generate a replicable series of variates, first run `set.seed(seedval)` where `seedval` is a single integer for the default "Mersenne-Twister" random number generator. For example:

R

```
set.seed(42)
set.seed(Sys.time())
```

Note: More information can be found using `help(.Random.seed)`.

1.11 Control flow, programming, and data generation

Programming is an area where SAS and R are quite different. Here we show some basic aspects of programming. We include parallel code for each language, while noting that some actions have no straightforward analogue in the other language.

1.11.1 Looping

SAS *HELP example:* see 6.1.2

```
data;
   do i = i1 to i2;
      x = normal(0);
      output;
   end;
run;
```

Note: The above code generates a new dataset with $i_2 - i_1 + 1$ standard Normal variates, with seed based on the system clock (1.10.5). The generic syntax for looping includes three parts: 1) a `do varname = val1 to val2` statement; 2) the statements to be executed within the loop; 3) an `end` statement. As with all programming languages, users should be careful about modifying the index during processing. Other options include `do while` and `do until`. To step values of `i` by values other than 1, use statements such as `do i = i1 to i2 by byval`. To step across specified values, use statements like `do k1, ... , kn`.

R

```
x <- numeric(i2-i1+1)   # create placeholder
for (i in 1:length(x)) {
   x[i] <- rnorm(1) # this is slow and inefficient!
}
```

or (preferably)

```
x <- rnorm(i2-i1+1)  # this is far better
```

Note: Most tasks in R that could be written as a loop are often dramatically faster if they are encoded as a vector operation (as in the second and preferred option above). Examples of situations where loops in R are particularly useful can be found in sections 3.1.6 and 6.1.2. More information on control structures for looping and conditional processing can be found in `help(Control)`.

1.11.2 Conditional execution

SAS *HELP example:* see 1.13.3 (SAS), 3.7.5 and 5.6.6 (R)

```
data ds;
   if expression1 then expression2 else expression3;
run;
```

or

```
if expression1 then expression2;
else if expression3 then expression4;
...
else expressionk;
```

or

```
if expression1 then do;
    ...;
    end;
else if expression2 then expression3;
...
```

Note: There is no limit on the number of conditions tested in the `else` statements, which always refer back to the most recent `if` statement. Once a condition in this sequence is met, the remaining conditions are not tested. Listing conditions in decreasing order of occurrence will therefore result more efficient code.

The `then` code is executed if the `expression` following the `if` has a nonmissing, nonzero value. So, for example, the statement `if 1 then y = x**2` is valid syntax, equivalent to the statement `y=x**2`. Good programming style is to make each tested expression be a logical test, such as `x eq 1` returning 1 if the expression is true and 0 otherwise. SAS includes mnemonics `lt`, `le`, `eq`, `ge`, `gt`, and `ne` for $<$, $\leq$, $=$, $\geq$, $>$, and $\neq$, respectively. The mnemonic syntax cannot be used for assignment, and it is recommended style to reserve `=` for assignment and use only the mnemonics for testing.

The `do-end` block is the equivalent of `{ }` in the R code below. Any group of `data` step statements can be included in a `do-end` block.

R

```
if (expression1) { expression2 }
```

or

```
if (expression1) { expression2 } else { expression3 }
```

or

```
ifelse(expression, x, y)
```

Note: The `if` statement, with or without `else`, tests a single logical statement; it is not an elementwise (vector) function. If `expression1` evaluates to `TRUE`, then `expression2` is evaluated. The `ifelse()` function operates on vectors and evaluates the expression given as `expression` and returns `x` if it is `TRUE` and `y` otherwise (see also comparisons, B.4.2). An expression can include multi-command blocks of code (in brackets).

1.11.3 Sequence of values or patterns

HELP example: see 1.13.5

It is often useful to generate a variable consisting of a sequence of values (e.g., the integers from 1 to 100) or a pattern of values (1 1 1 2 2 2 3 3 3). This might be needed to generate a variable consisting of a set of repeated values for use in a simulation or graphical display.

As an example, we demonstrate generating data from a linear regression model of the form:

$$E[Y|X_1, X_2] = \beta_0 + \beta_1 X_1 + \beta_2 X_2,\ Var(Y|X) = 3,\ Corr(X_1, X_2) = 0.$$

SAS

```
data ds;
   do x = 1 to nvals;
   ...
   end;
run;
```

Note: The following code implements the model described above for $n = 200$. The value 42 below is an arbitrary seed (1.10.9) [1] used for random number generation. The datasets ds1 and ds2 will be identical. However such values are generated, it would be wise to use `proc freq` (2.3.1) to check whether the intended results were achieved.

```
data ds1;
   beta0 = -1; beta1 = 1.5; beta2 = .5; mse = 3;
      /* note multiple statements on previous line */
   do x1 = 1 to 2;
      do x2 = 1 to 2;
         do obs = 1 to 50;
            y = beta0 + beta1*x1 + beta2*x2 + normal(42)*mse;
            output;
          end;
      end;
   end;
run;
```

or

```
data ds2;
   beta0 = -1; beta1 = 1.5; beta2 = .5; mse = 3;
   do i = 1 to 200;
      x1 = (i gt 100) + 1;
      x2 = (((i gt 50) and (i le 100)) or (i gt 150))  + 1;
      y = beta0 + beta1*x1 + beta2*x2 + normal(42)*mse;
      output;
   end;
run;
```

R

```
# generate
seq(from=i1, to=i2, length.out=nvals)
seq(from=i1, to=i2, by=1
seq(i1, i2)
i1:i2

rep(value, times=nvals)
```

or

```
rep(value, each=nvals)
```

Note: The `seq` function creates a vector of length `val` if the `length.out` option is specified. If the `by` option is included, the length is approximately `(i2-i1)/byval`. The `i1:i2` operator is equivalent to `seq(from=i1, to=i2, by=1)`. The `rep` function creates a vector of length `nvals` with all values equal to `value`, which can be a scalar, vector, or list. The `each` option repeats each element of `value nvals` times. The default is `times`.

The following code implements the model described above for $n = 200$.

```
> n <- 200
> x1 <- rep(c(0,1), each=n/2)      # x1 resembles 0 0 0 ... 1 1 1
> x2 <- rep(c(0,1), n/2)           # x2 resembles 0 1 0 1 ... 0 1
> beta0 <- -1; beta1 <- 1.5; beta2 <- .5;
> mse <- 3
> table(x1, x2)
   x2
x1   0  1
  0 50 50
  1 50 50
> y <- beta0 + beta1*x1 + beta2*x2 + rnorm(n, 0, mse)
> lm(y ~ x1 + x2)
```

1.11.4 Referring to a range of variables

HELP example: see 1.13.3

For functions such as `mean()` it is often desirable to list variables to be averaged without listing them all by name. SAS provides two ways of doing this. First, variables stored adjacently can be referred to as a range `vara -- varb` (with two hyphens). Variables with sequential numerical suffices can be referred to as a range `varname1 - varnamek` (with a single hyphen) regardless of the storage location. The key thing to bear in mind is that the part of the name before the number must be identical for all variables. This shorthand syntax also works in procedures.

No straightforward equivalent exists in R, though variables stored adjacently in a dataframe can be referred using indexing by their column number.

SAS

```
data ...;
   meanadjacentx = mean(of x1 -- xk);
   meannamedx = mean(of x1 - xk);
run;
```

Note: The former code will return the mean of all the variables stored between x_1 and x_k. The latter will return the mean of $x_1 \ldots x_k$, if they all exist.

1.11.5 Perform an action repeatedly over a set of variables

HELP example: see 1.13.3, 4.6.9

It is often necessary to perform a given function for a series of variables. Here the square of each of a list of variables is calculated as an example.

In SAS, this can be accomplished using arrays.

SAS

```
data ...;
   array arrayname1 [arraylen] x1 x2 ... xk;
   array arrayname2 [arraylen] z1 ... zk;
   do i = 1 to arraylen;
      arrayname2[i] = arrayname1[i]**2;
   end;
run;
```

Note: In the above example, $z_i = x_i^2, i = 1 \ldots k$, for every observation in the dataset. The variable `arraylen` is an integer. It can be replaced by '*', which implies that the dimension of the array is to be calculated automatically by SAS from the number of elements. Elements (variables in the array) are listed after the brackets. Arrays can also be multidimensional, when multiple dimensions are specified (separated by commas) within the brackets. This can be useful, for example, when variables contain a matrix for each observation in the dataset.

Variables can be created by definition in the array statement, meaning that in the above code, the variable `x2` need not exist prior to the first `array` statement. The function `dim(arrayname1)` returns the number of elements in the array, and can be used in place of the variable `arraylen` to loop over arrays declared with the '*' syntax.

R

```
l1 <- c("x1", "x2", ..., "xk")
l2 <- c("z1", "z2", ..., "zk")
for (i in 1:length(l1)) {
    assign(l2[i], eval(as.name(l1[i]))^2)
}
```

Note: It is not straightforward to refer to objects within R without evaluating those objects. Assignments to R objects given symbolically can be made using the `assign()` function. Here a nonobvious use of the `eval()` function is used to evaluate an expression after the string value in `l1` is coerced to be a symbol. This allows the values of the character vectors `l1` and `l2` to be evaluated (see `help(assign)` and `help(eval)`).

1.12 Further resources

Comprehensive introductions to data management in SAS can be found in [15] and [9]. Similar developments in R are accessibly presented in [95]. Paul Murrell's forthcoming *Introduction to Data Technologies* text [57] provides a comprehensive introduction to XML, SQL, and other related technologies and can be found at `http://www.stat.auckland.ac.nz/~paul/ItDT`.

1.13 HELP examples

To help illustrate the tools presented in this chapter, we apply many of the entries to the HELP data. SAS and R code can be downloaded from `http://www.math.smith.edu/sasr/examples`.

1.13.1 Data input and output

We begin by reading the dataset (1.1.4), keeping only the variables that are needed (1.5.8).

```
proc import
   datafile='c:/book/help.csv'
   out=dsprelim
   dbms=dlm;
   delimiter=',';
   getnames=yes;
run;

data ds;
set dsprelim;
   keep id cesd f1a -- f1t i1 i2 female treat;
run;
```

```
> options(digits=3)
> options(width=72) # narrow output
> ds <- read.csv("http://www.math.smith.edu/sasr/datasets/help.csv")
> newds <- ds[,c("cesd","female","i1","i2","id","treat","f1a","f1b",
+     "f1c","f1d","f1e","f1f","f1g","f1h","f1i","f1j","f1k","f1l","f1m",
+     "f1n","f1o","f1p","f1q","f1r","f1s","f1t")]
```

We can then show a summary of the dataset. In SAS, we use the ODS system (A.7) to reduce the length of the output.

```
options ls=74;   /* narrows width to stay in grey box */
ods select attributes;
proc contents data=ds;
run;
ods select all;

The CONTENTS Procedure
Data Set Name          WORK.DS                   Observations          453
Member Type            DATA                      Variables             26
Engine                 V9                        Indexes               0
Created                Tuesday, March 10,        Observation Length    208
                       2009 04:50:43 PM
Last Modified          Tuesday, March 10,        Deleted Observations  0
                       2009 04:50:43 PM
Protection                                       Compressed            NO
Data Set Type                                    Sorted                NO
Label
Data Representation    WINDOWS_32
Encoding               wlatin1  Western (Windows)
```

The default output prints a line for each variable with its name and additional information; the short option below limits the output to just the names of the variable.

```
options ls=74;   /* narrows width to stay in grey box */
ods select variablesshort;
proc contents data=ds short;
run;
ods select all;

The CONTENTS Procedure
                 Alphabetic List of Variables for WORK.DS
cesd f1a f1b f1c f1d f1e f1f f1g f1h f1i f1j f1k f1l f1m f1n f1o f1p f1q
f1r f1s f1t female i1 i2 id treat
```

```
> attach(newds)
> names(newds)

 [1] "cesd"   "female" "i1"     "i2"     "id"     "treat"  "f1a"
 [8] "f1b"    "f1c"    "f1d"    "f1e"    "f1f"    "f1g"    "f1h"
[15] "f1i"    "f1j"    "f1k"    "f1l"    "f1m"    "f1n"    "f1o"
[22] "f1p"    "f1q"    "f1r"    "f1s"    "f1t"

> # structure of the first 10 variables
> str(newds[,1:10])

'data.frame':        453 obs. of  10 variables:
 $ cesd  : int  49 30 39 15 39 6 52 32 50 46 ...
 $ female: int  0 0 0 1 0 1 1 0 1 0 ...
 $ i1    : int  13 56 0 5 10 4 13 12 71 20 ...
 $ i2    : int  26 62 0 5 13 4 20 24 129 27 ...
 $ id    : int  1 2 3 4 5 6 7 8 9 10 ...
 $ treat : int  1 1 0 0 0 1 0 1 0 1 ...
 $ f1a   : int  3 3 3 0 3 1 3 1 3 2 ...
 $ f1b   : int  2 2 2 0 0 0 1 1 2 3 ...
 $ f1c   : int  3 0 3 1 3 1 3 2 3 3 ...
 $ f1d   : int  0 3 0 3 3 3 1 3 1 0 ...
```

Displaying the first few rows of data can give a more concrete sense of what is in the dataset:

```
proc print data=ds (obs=5) width=minimum;
run;

                                                        f
                                                   t    e
                                                   r  c m
 O   f f f f f f f f f f f f f f f f f f f f       e  e a
 b i 1 1 1 1 1 1 1 1 1 1 1 1 1 1 1 1 1 1 1 1  i  i a  s l
 s d a b c d e f g h i j k l m n o p q r s t  1  2 t  d e
  1 1 3 2 3 0 2 3 3 0 2 3 3 0 1 2 2 2 2 3 3 2 13 26 1 49 0
  2 2 3 2 0 3 3 2 0 0 3 0 3 0 0 3 0 0 0 2 0 0 56 62 1 30 0
  3 3 3 2 3 0 2 2 1 3 2 3 1 0 1 3 2 0 0 3 2 0  0  0 0 39 0
  4 4 0 0 1 3 2 2 1 3 0 0 1 2 2 2 0 . 2 0 0 1  5  5 0 15 1
  5 5 3 0 3 3 3 3 1 3 3 2 3 2 2 3 0 3 3 3 3 3 10 13 0 39 0
```

```
> head(newds, n=5)

  cesd female i1 i2 id treat f1a f1b f1c f1d f1e f1f f1g f1h f1i f1j
1   49      0 13 26  1     1   3   2   3   0   2   3   3   0   2   3
2   30      0 56 62  2     1   3   2   0   3   3   2   0   0   3   0
3   39      0  0  0  3     0   3   2   3   0   2   2   1   3   2   3
4   15      1  5  5  4     0   0   0   1   3   2   2   1   3   0   0
5   39      0 10 13  5     0   3   0   3   3   3   3   1   3   3   2
  f1k f1l f1m f1n f1o f1p f1q f1r f1s f1t
1   3   0   1   2   2   2   2   3   3   2
2   3   0   0   3   0   0   0   2   0   0
3   1   0   1   3   2   0   0   3   2   0
4   1   2   2   2   0  NA   2   0   0   1
5   3   2   2   3   0   3   3   3   3   3
```

Saving the dataset in native format (1.2.1) will ease future access. We also add a comment (1.3.5) to help later users understand what is in the dataset.

```
libname book 'c:/temp';
data book.ds (label = "HELP baseline dataset");
set ds;
run;
```

```
> comment(newds) <- "HELP baseline dataset"
> comment(newds)

[1] "HELP baseline dataset"

> save(ds, file="savedfile")
```

Saving it in a foreign format (1.1.5), say Microsoft Excel, will allow access to other tools for analysis and display:

```
proc export data=ds replace
   outfile="c:/temp/ds.xls"
   dbms=excel;
run;
```

Getting data into SAS format from R is particularly useful; note that the R code below generates an ASCII dataset and a SAS command file to read it in to SAS.

```
> library(foreign)
> write.foreign(newds, "file.dat", "file.sas", package="SAS")
```

1.13.2 Data display

We begin by consideration of the CESD (Center for Epidemiologic Statistics) measure of depressive symptoms for this sample at baseline.

```
proc print data=ds (obs=10);
   var cesd;
run;

Obs           cesd
  1             49
  2             30
  3             39
  4             15
  5             39
  6              6
  7             52
  8             32
  9             50
 10             46
```

The indexing mechanisms in R (see B.4.2) are helpful in extracting subsets of a vector.

```
> cesd[1:10]

 [1] 49 30 39 15 39  6 52 32 50 46
```

It may be useful to know how many high values there are, and to which observations they belong:

```
proc print data=ds;
   where cesd gt 55;
   var cesd;
run;

Obs          cesd
 64            57
116            58
171            57
194            60
231            58
266            56
295            58
305            56
387            57
415            56
```

```
> cesd[cesd>55]

 [1] 57 58 57 60 58 56 58 56 57 56

> # which rows have values this high?
> which(cesd>55)

 [1]  64 116 171 194 231 266 295 305 387 415
```

Similarly, it may be useful to examine the observations with the lowest values:

```
proc sort data=ds out=dss1;
   by cesd;
run;

proc print data=dss1 (obs=4);
   var id cesd i1 treat;
run;

Obs          id          cesd          i1          treat
  1         233             1           3              0
  2         418             3          13              0
  3         139             3           1              0
  4          95             4           9              1
```

```
> sort(cesd)[1:4]

[1] 1 3 3 4
```

1.13.3 Derived variables and data manipulation

Suppose the dataset arrived with only the individual CESD questions, and not the sum. We would need to create the CESD score. In SAS, we'll do this using an array (1.11.5) to aid the recoding of the four questions which are asked "backwards," meaning that high values of the response are counted for fewer points.[1] In R we'll approach the backwards questions by reading the CESD items into a new object. To demonstrate other tools, we'll also see if there's any missing data (1.4.14), and how the original creators of the dataset handled it.

[1]according to the coding instructions at `http://patienteducation.stanford.edu/research/cesd.pdf`

```
data cesd;
set ds;
   /* list of backwards questions */
   array backwards [*] f1d f1h f1l f1p;
   /* for each, subtract the stored value from 3 */
   do i = 1 to dim(backwards);
      backwards[i] = 3 - backwards[i];
   end;
   /* this generates the sum of the non-missing questions */
   newcesd = sum(of f1a -- f1t);
   /* This counts the number of missing values, per person */
   nmisscesd = nmiss(of f1a -- f1t);
   /* this gives the sum, imputing the mean of non-missing */
   imputemeancesd = mean(of f1a -- f1t) * 20;
run;
```

```
> table(is.na(f1g))

FALSE  TRUE
  452     1

> # reverse code f1d, f1h, f1l and f1p
> cesditems <- cbind(f1a, f1b, f1c, (3 - f1d), f1e, f1f, f1g,
+     (3 - f1h), f1i, f1j, f1k, (3 - f1l), f1m, f1n, f1o, (3 - f1p),
+     f1q, f1r, f1s, f1t)
> nmisscesd <- apply(is.na(cesditems), 1, sum)
> ncesditems <- cesditems
> ncesditems[is.na(cesditems)] <- 0
> newcesd <- apply(ncesditems, 1, sum)
> imputemeancesd <- 20/(20-nmisscesd)*newcesd
```

It is prudent to review the results when deriving variables. We'll check our recreated CESD score against the one which came with the dataset. To ensure that missing data has been correctly coded, we print the subjects with any missing questions.

```
proc print data=cesd (obs=20);
   where nmisscesd gt 0;
   var cesd newcesd nmisscesd imputemeancesd;
run;

Obs          cesd     newcesd     nmisscesd     imputemeancesd
  4            15          15             1            15.7895
 17            19          19             1            20.0000
 87            44          44             1            46.3158
101            17          17             1            17.8947
154            29          29             1            30.5263
177            44          44             1            46.3158
229            39          39             1            41.0526
```

```
> cbind(newcesd, cesd, nmisscesd, imputemeancesd)[nmisscesd>0,]

     newcesd cesd nmisscesd imputemeancesd
[1,]      15   15         1           15.8
[2,]      19   19         1           20.0
[3,]      44   44         1           46.3
[4,]      17   17         1           17.9
[5,]      29   29         1           30.5
[6,]      44   44         1           46.3
[7,]      39   39         1           41.1
```

The output shows that the original dataset was created with unanswered questions counted as if they had been answered with a zero. This conforms to the instructions provided with the CESD, but might be questioned on theoretical grounds.

It is often necessary to create a new variable using logic (1.4.11). In the HELP study, many subjects reported extreme amounts of drinking (as the baseline measure was taken while they were in detox). Here, an ordinal measure of alcohol consumption (abstinent, moderate, high-risk) is created using information about average consumption per day in past 30 days prior to detox (`i1`, measured in standard drink units) and maximum number of drinks per day in past 30 days prior to detox (`i2`). The number of drinks required for each category differ for men and women according to NIAAA guidelines for physicians [59].

```
data ds2;
set ds;
   if i1 eq 0 then drinkstat="abstinent";
   if (i1 eq 1 and i2 le 3 and female eq 1) or
   (((i1 eq 1) or (i1 eq 2)) and i2 le 4 and female eq 0)
      then drinkstat="moderate";
   if (((i1 gt 1) or (i2 gt 3)) and female eq 1) or
   (((i1 gt 2) or (i2 gt 4)) and female eq 0)
      then drinkstat="highrisk";
   if nmiss(i1,i2,female) ne 0 then drinkstat="";
run;
```

```
> # create empty repository for new variable
> drinkstat <- character(length(i1))
> # create abstinent group
> drinkstat[i1==0] <- "abstinent"
> # create moderate group
> drinkstat[(i1>0 & i1<=1 & i2<=3 & female==1) |
+   (i1>0 & i1<=2 & i2<=4 & female==0)] <- "moderate"
> # create highrisk group
> drinkstat[((i1>1 | i2>3) & female==1) |
+   ((i1>2 | i2>4) & female==0)] <- "highrisk"
> # do we need to account for missing values?
> is.na(drinkstat) <- is.na(i1) | is.na(i2) | is.na(female)
> table(is.na(drinkstat))

FALSE
  453
```

It is always prudent to check the results of derived variables. As a demonstration, we display the observations in the 361st through 370th rows of the data.

```
proc print data=ds2 (firstobs=361 obs=370);
   var i1 i2 female drinkstat;
run;

Obs              i1              i2          female    drinkstat
361              37              37               0    highrisk
362              25              25               0    highrisk
363              38              38               0    highrisk
364              12              29               0    highrisk
365               6              24               0    highrisk
366               6               6               0    highrisk
367               0               0               0    abstinent
368               0               0               1    abstinent
369               8               8               0    highrisk
370              32              32               0    highrisk
```

```
> tmpds <- data.frame(i1, i2, female, drinkstat)
> tmpds[361:370,]

    i1 i2 female drinkstat
361 37 37      0  highrisk
362 25 25      0  highrisk
363 38 38      0  highrisk
364 12 29      0  highrisk
365  6 24      0  highrisk
366  6  6      0  highrisk
367  0  0      0 abstinent
368  0  0      1 abstinent
369  8  8      0  highrisk
370 32 32      0  highrisk
```

It is also useful to focus such checks on a subset of observations. Here we show the drinking data for moderate female drinkers.

```
proc print data=ds2;
   where drinkstat eq "moderate" and female eq 1;
   var i1 i2 female drinkstat;
run;

Obs              i1              i2          female    drinkstat
116               1               1               1    moderate
137               1               3               1    moderate
225               1               2               1    moderate
230               1               1               1    moderate
264               1               1               1    moderate
266               1               1               1    moderate
394               1               1               1    moderate
```

```
> tmpds[tmpds$drinkstat=="moderate" & tmpds$female==1,]

    i1 i2 female drinkstat
116  1  1      1  moderate
137  1  3      1  moderate
225  1  2      1  moderate
230  1  1      1  moderate
264  1  1      1  moderate
266  1  1      1  moderate
394  1  1      1  moderate
```

Basic data description is an early step in analysis. Here we show some summary statistics related to drinking and gender.

```
proc freq data=ds2;
   tables drinkstat;
run;

The FREQ Procedure
                                              Cumulative    Cumulative
drinkstat    Frequency      Percent     Frequency       Percent
---------------------------------------------------------------
abstinent           68        15.01            68         15.01
highrisk           357        78.81           425         93.82
moderate            28         6.18           453        100.00
```

```
> sum(is.na(drinkstat))

[1] 0

> table(drinkstat, exclude="NULL")

drinkstat
abstinent  highrisk  moderate
       68       357        28
```

```
proc freq data=ds2;
   tables drinkstat*female;
run;
```

```
The FREQ Procedure
Table of drinkstat by female
drinkstat     female
Frequency |
Percent   |
Row Pct   |

Col Pct   |       0|       1|  Total
----------+--------+--------+
abstinent |     42 |     26 |     68
          |   9.27 |   5.74 |  15.01
          |  61.76 |  38.24 |
          |  12.14 |  24.30 |
----------+--------+--------+
highrisk  |    283 |     74 |    357
          |  62.47 |  16.34 |  78.81
          |  79.27 |  20.73 |
          |  81.79 |  69.16 |
----------+--------+--------+
moderate  |     21 |      7 |     28
          |   4.64 |   1.55 |   6.18
          |  75.00 |  25.00 |
          |   6.07 |   6.54 |
----------+--------+--------+
Total          346      107      453
             76.38    23.62   100.00
```

```
> table(drinkstat, female, exclude="NULL")

           female
drinkstat     0   1
  abstinent  42  26
  highrisk  283  74
  moderate   21   7
```

To display gender in a more direct fashion, we create a new character variable. Note that in these quoted strings, both SAS and R are case sensitive.

```
data ds3;
set ds;
   if female eq 1 then gender="Female";
   else if female eq 0 then gender="male";
run;

proc freq data=ds3;
   tables female gender;
run;
```

```
The FREQ Procedure
                                    Cumulative    Cumulative
female    Frequency     Percent     Frequency      Percent
-----------------------------------------------------------
     0         346       76.38           346        76.38
     1         107       23.62           453       100.00

                                    Cumulative    Cumulative
gender    Frequency     Percent     Frequency      Percent
-----------------------------------------------------------
Female         107       23.62           107        23.62
male           346       76.38           453       100.00
```

```
> gender <- factor(female, c(0,1), c("male","Female"))
> table(female)

female
  0   1
346 107

> table(gender)

gender
  male Female
   346    107
```

1.13.4 Sorting and subsetting datasets

It is often useful to sort datasets (1.5.6) by the order of a particular variable (or variables). Here we sort by CESD and drinking.

```
proc sort data=ds;
   by cesd i1;
run;

proc print data=ds (obs=5);
   var id cesd i1;
run;

Obs             id              cesd              i1
  1            233                 1               3
  2            139                 3               1
  3            418                 3              13
  4            251                 4               4
  5             95                 4               9
```

```
> detach(newds)
> ds <- read.csv("help.csv")
> newds <- ds[order(ds$cesd, ds$i1),]
> newds[1:5,c("cesd", "i1", "id")]

    cesd i1  id
199    1  3 233
394    3  1 139
349    3 13 418
417    4  4 251
85     4  9  95
```

It is sometimes necessary to create data that is a subset (1.5.1) of other data. For example, here we make a dataset which only includes female subjects. First, we create the subset and calculate a summary value in the resulting dataset.

```
data females;
set ds;
   where female eq 1;
run;

proc means data=females mean maxdec=1;
  var cesd;
run;

The MEANS Procedure
Analysis Variable : cesd

        Mean
------------
        36.9
------------
```

```
> females <- ds[ds$female==1,]
> attach(females)
> mean(cesd)

[1] 36.9
```

To test the subsetting, we then display the mean for both genders.

```
proc sort data=ds;
   by female;
run;

proc means data=ds mean maxdec=2;
   by female;
   var cesd;
run;

female=0
The MEANS Procedure
Analysis Variable : cesd

        Mean
------------
       31.60
------------

female=1
Analysis Variable : cesd

        Mean
------------
       36.89
------------
```

```
> ds <- read.csv("help.csv")
> tapply(ds$cesd, ds$female, mean)

   0    1
31.6 36.9
```

1.13.5 Probability distributions

To demonstrate more tools, we leave the HELP dataset and show examples of how data can be generated within each programming environment. We will generate values (1.10.5) from the normal and t distribution densities; note that the probability density functions are not hard-coded into SAS as they are within R.

```
data dists;
   do x = -4 to 4 by .1;
      normal_01 = sqrt(2 * constant('PI'))**(-1) * exp(-1 * ((x*x)/2)) ;
      dfval = 1;
      t_1df = (gamma((dfval +1)/2) / (sqrt(dfval * constant('PI')) *
         gamma(dfval/2))) * (1 + (x*x)/dfval)**(-1 * ((dfval + 1)/2));
      output;
      end;
run;
```

```
> x <- seq(from=-4, to=4.2, length=100)
> normval <- dnorm(x, 0, 1)
> dfval <- 1
> tval <- dt(x, df=dfval)
```

Figure 1.1 displays a plot of these distributions in SAS and R.

```
legend1 label=none position=(top inside right) frame down=2
   value = ("N(0,1)" tick=2 "t with 1 df");
axis1 label=(angle=90 "f(x)") minor=none order=(0 to .4 by .1);
axis2 minor=none order=(-4 to 4 by 2);
symbol1 i=j v=none l=1 c=black w=5;
symbol2 i=j v=none l=21 c=black w=5;
proc gplot data= dists;
   plot (normal_01 t_1df) * x / overlay legend=legend1
      vaxis=axis1 haxis=axis2;
run; quit;
```

```
> plot(x, normval, type="n", ylab="f(x)", las=1)
> lines(x, normval, lty=1, lwd=2)
> lines(x, tval, lty=2, lwd=2)
> legend(1.1, .395, lty=1:2, lwd=2,
+    legend=c(expression(N(mu == 0,sigma == 1)),
+    paste("t with ", dfval," df", sep="")))
```

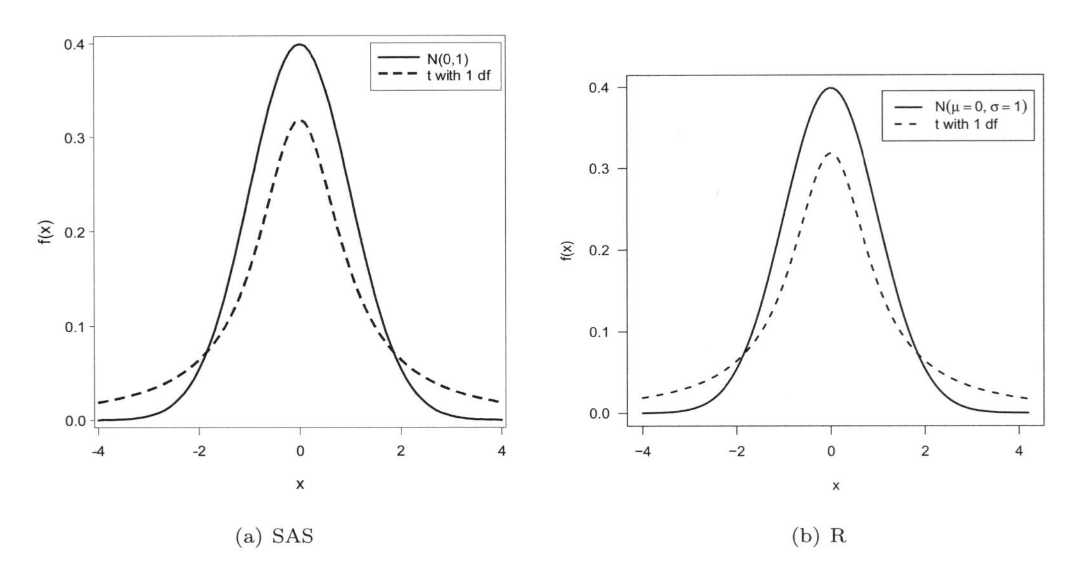

(a) SAS (b) R

Figure 1.1: Comparison of standard normal and t distribution with 1 df

Chapter 2

Common statistical procedures

This chapter describes how to generate univariate summary statistics for continuous variables (such as means, variances, and quantiles), display and analyze frequency tables and cross-tabulations for categorical variables, as well as carry out a variety of one and two sample procedures.

2.1 Summary statistics

2.1.1 Means and other summary statistics

SAS *HELP example:* see 2.6.1

```
proc means data=ds keyword1 ... keywordn;
   var x1 ... xk;
run;
```

or

```
proc summary data=ds;
   var x1 ... xk;
   output out=newds keyword1= keyword2(x2)=newname
      keyword3(x3 x4)=newnamea newnameb;
run;

proc print data=newds;
run;
```

or

```
proc univariate data=ds;
   var x1 ... xk;
run;
```

Note: The `univariate` procedure generates a number of statistics by default, including the mean, standard deviation, skewness, and kurtosis. The `means` and `summary` procedures accept a number of `keywords`, including `mean`, `median`, `var`, `stdev`, `min`, `max`, `sum`. These procedures are identical except that `proc summary` produces no printed output, only an output dataset, while `proc means` can produce both printed output and a dataset. The `output` statement syntax is `keyword=` in which case the summary statistic shares the name of the variable summarized, `keyword(varname)=newname` in which case the summary statistic takes the new name, or `keyword(varname1 ... varnamek)=newname1 ... newnamek` which allows the naming of many summary statistic variables at once. These options become valuable especially when summarizing within subgroups (2.1.2). The `maxdec` option to the `proc means` statement controls the number of decimal places printed.

R

```
xmean <- mean(x)
```

Note: The `mean()` function accepts a numeric vector or a numeric dataframe as arguments (date objects are also supported). Similar functions in R include `median()` (see 2.1.5 for more quantiles), `var()`, `sd()`, `min()`, `max()`, `sum()`, `prod()`, and `range()` (note that the latter returns a vector containing the minimum and maximum value).

2.1.2 Means by group

HELP example: see 2.6.4 and 1.13.4

SAS

```
proc sort data=ds;
   by y;
run;

proc means data=ds;
   by y;
   var x;
run;
```

or

```
proc sort data=ds;
   by y;
run;

proc summary data=ds;
   by y;
   output out=newds mean=;
   var x;
run;
proc print data=newds;
run;
```

Note: The summary statistics for each `by` group are included in any printed output and in any datasets created by the procedure. See section 2.1.1 for a discussion of `output` statement syntax.

R

```
tapply(x, y, mean)
```

or

```
ave(x, as.factor(y), FUN=mean)
```

Note: The `tapply()` function applies the specified function given as the third argument (in this case `mean()`) to the vector `y` stratified by every unique set of values of the list of factors specified `x`. It returns a vector with length equal to the number of unique set of values of `x`. Similar functionality is available using the `ave()` function (see `example(ave)`), which returns a vector of the same length as `x` with each element equal to the mean of the subset of observations with the factor level specified by `y`.

2.1.3 Trimmed mean

SAS

```
proc univariate data=ds trimmed=frac;
   var x;
run;
```

Note: The parameter `frac` is the proportion of observations above and below the mean to exclude, or a number (greater than 1) in which case `number` observations will be excluded. Multiple variables may be specified. This statistic can be saved into a dataset using `ODS` (see A.7).

R

```
mean(x, trim=frac)
```

Note: The value `frac` can take on range 0 to 0.5, and specifies the fraction of observations to be trimmed from each end of `x` before the mean is computed (`frac=0.5` yields the median).

2.1.4 Five-number summary

HELP example: see 2.6.1

The five number summary (minimum, 25th percentile, median, 75th percentile, maximum) is a useful summary of the distribution of observed values.

SAS

```
proc means data=ds mean min q1 median q3 max;
   var x1 ... xk;
run;
```

R

```
quantile(x)
fivenum(x)
summary(ds)
```

Note: The `summary()` function calculates the five number summary (plus the mean) for each of the columns of the vector or dataset given as arguments. The default output of the `quantile()` function is the min, 25th percentile, median, 75th percentile and the maximum. The `fivenum()` function reports the lower and upper hinges instead of the 25th and 75th percentiles, respectively.

2.1.5 Quantiles

SAS

HELP example: see 2.6.1

```
proc univariate data=ds;
   var x1 ... xk;
   output out=newds pctlpts=2.5, 95 to 97.5 by 1.25
      pctlpre=p pctlnames=2_5 95 96_125 97_5;
run;
```

Note: This creates a new dataset with the 2.5, 95, 96.25, 97.5 values stored in variables named p2_5, p95, p96_125, and p97_5. The first, 5th, 10th, 25th, 50th, 75th, 90th, 95th, and 99th can be obtained more directly from `proc means`, `proc summary`, and `proc univariate`.

Details and options regarding calculation of quantiles in `proc univariate` can be found in SAS on-line help: Contents; SAS Products; SAS Procedures; UNIVARIATE; Calculating Percentiles.

R

```
quantile(x, c(.025, .975))
quantile(x, seq(from=.95, to=.975, by=.0125))
```

Note: Details regarding the calculation of quantiles in `quantile()` can be found using `help(quantile)`.

2.1.6 Centering, normalizing, and scaling

SAS

```
proc standard data=ds out=ds2 mean=0 std=1;
   var x1 ... xk;
run;
```

Note: The output dataset named in the `out` option contains all of the data from the original dataset, with the standardized version of each variable named in the `var` statement stored in place of the original. Either the `mean` or the `std` option may be omitted.

R

```
zscoredx <- scale(x)
```

or

```
zscoredx <- (x-mean(x))/sd(x)
```

Note: The default behavior of `scale()` is to create a Z-score transformation. The `scale()` function can operate on matrices and dataframes, and allows the specification of a vector of the scaling parameters for both center and scale (see also `sweep()`, a more general function).

2.1.7 Mean and 95% confidence interval

SAS

```
proc means data=ds lclm mean uclm;
   var x;
run;
```

Note: Calculated statistics can be saved using an `output` statement or using `proc summary` as in 2.1.1 or using `ODS`.

R

```
tcrit <- qt(.975, length(x)-1)
ci95 <- c(mean(x) - tcrit*sd(x)/sqrt(length(x)),
          mean(x) + tcrit*sd(x)/sqrt(length(x)))
```

or

```
t.test(x)$conf.int
```

Note: While the appropriate 95% confidence interval can be generated in terms of the mean and standard deviation, it is more straightforward to use the t-test function to calculate the relevant quantities.

2.1.8 Bootstrapping a sample statistic

Bootstrapping is a powerful and elegant approach to estimation of sample statistics that can be implemented even in many situations where asymptotic results are difficult to find or otherwise unsatisfactory [18]. Bootstrapping proceeds using three steps: first, resample the dataset (with replacement) a specified number of times (typically on the order of 10000), calculate the desired statistic from each resampled dataset, then use the distribution of the resampled statistics to estimate the standard error of the statistic (normal approximation method), or construct a confidence interval using quantiles of that distribution (percentile method).

As an example, we consider estimating the standard error and 95% confidence interval for the coefficient of variation (COV), defined as σ/μ, for a random variable X. Note that for both packages, the user must provide code to calculate the statistic of interest; this must be done in a macro (in SAS), or as a function (in R).

SAS

```
/* download "jackboot.sas" from http://support.sas.com/kb/24/982.html */
%include 'c:/sasmacros/jackboot.sas';
/* create macro that generates the desired statistic, in this case the
   coefficient of variation, just once, from the observed data.
   This macro must be named %analyze  */
%macro analyze(data=, out=);
proc summary data=&data;
   var x;
   output out=&out (drop=_freq_ _type_) cv=cv_x;
run;
%mend;

%boot(data=ds, samples=1000);
```

Note: The `%include` statement is equivalent to typing the contents of the included file into the program. The `%boot` macro requires an existing `%analyze` macro, which must generate an output dataset; bootstrap results for all variables in this output dataset are calculated. The `drop` data set option removes some character variables from this output dataset so that statistics are not reported on them. See section A.8 for more information on SAS macros.

R

```
library(boot)
covfun <- function(x, i) {sd(x[i])/mean(x[i])}
res <- boot(x, covfun, R=10000)
summary(res)
plot(res)
quantile(res$t, c(.025, .975))
mean(res$t) + c(-1.96, 1.96)*sd(res$t)
```

Note: The first argument to the `boot()` function specifies the data to be bootstrapped (in this case a vector, though a dataframe can be set up if more than one variable is needed for the calculation of the sample statistic) as well as a function to calculate the statistic for each resampling iteration. Here the function `covfun()` takes two arguments: the first is the original data (as a vector) and the second a set of indices into that vector (that represent a given bootstrap sample).

The `boot()` function returns an object of class `boot`, with an associated `plot()` function that provides a histogram and QQ-plot (see `help(plot.boot)`). The return value object (`res`, above) contains the vector of resampled statistics (`res$t`), which can be used to estimate the standard error and 95% confidence interval. The `boot.ci()` function can be used to generate bias-corrected and accelerated intervals.

2.1.9 Proportion and 95% confidence interval

HELP example: see 6.1.2

SAS

```
proc freq data=ds;
   tables x / binomial;
run;
```

Note: The binomial option requests the exact Clopper–Pearson confidence interval based on the F distribution [10], an approximate confidence interval, and a test that the probability of the first level of the variable = 0.5. If x has more than two levels, the probability estimated and tested is the probability of the first level vs. all the others combined. Additional confidence intervals are available as options to the `binomial` option.

R

```
binom.test(sum(x), length(x))
prop.test(sum(x), length(x))
```

Note: The `binom.test()` function calculates an exact Clopper–Pearson confidence interval based on the F distribution [10] using the first argument as the number of successes and the second argument the number of trials, while `prop.test()` calculates an approximate confidence interval by inverting the score test. Both allow specification of p for the null hypothesis. The `conf.level` option can be used to change the default confidence level.

2.2 Bivariate statistics

2.2.1 Epidemiologic statistics

HELP example: see 2.6.3

SAS

```
proc freq data=ds;
   tables x*y / relrisk;
run;
```

R

```
sum(x==0&y==0)*sum(x==1&y==1)/(sum(x==0&y==1)*sum(x==1&y==0))
```

or

```
tab1 <- table(x, y)
tab1[1,1]*tab1[2,2]/(tab1[1,2]*tab1[2,1])
```

or

```
glm1 <- glm(y ~ x, family=binomial)
exp(glm1$coef[2])
```

or

```
library(epitools)
oddsratio.fisher(x, y)
oddsratio.wald(x, y)
riskratio(x, y)
riskratio.wald(x, y)
```

Note: The `epitab()` function in `library(epitools)` provides a general interface to many epidemiologic statistics, while `expand.table()` can be used to create individual level data from a table of counts (see also generalized linear models, 4.1).

2.2.2 Test characteristics

The sensitivity of a test is defined as the probability that someone with the disease (D=1) tests positive (T=1), while the specificity is the probability that someone without the disease (D=0) tests negative (T=0). For a dichotomous screening measure, the sensitivity and specificity can be defined as $P(D = 1, T = 1)/P(D = 1)$ and $P(D = 0, T = 0)/P(D = 0)$, respectively. (See also receiver operating character curves, 5.1.18.)

SAS

```
proc freq data=ds;
   tables d*t / out=newds;
run;

proc means data=newds nway;
   by d;
   var count;
   output out=newds2 sum=sumdlev;
run;

data newds3;
merge newds newds2;
   by d;
   retain sens spec;
   if D eq 1 and T=1 then sens=count/sumdlev;
   if D eq 0 and T=0 then spec=count/sumdlev;
   if sens ge 0 and spec ge 0;
run;
```

Note: The above code creates a dataset with a single line containing the sensitivity, specificity, and other data, given a test positive indicator `t` and disease indicator `d`. Sensitivity and specificity across all unique cut-points of a continuous measure `T` can be calculated as follows.

```
proc summary data=ds;
   var d;
   output out=sumdisease sum=totaldisease n=totalobs;
run;

proc sort data=ds; by descending t; run;
```

```
data ds2;
set ds;
   if _n_ eq 1 then set sumdisease;
   retain sumdplus 0 sumdminus 0;
   sumdplus = sumdplus + d;
   sumdminus = sumdminus + (d eq 0);
   sens = sumdplus/totaldisease;
   one_m_spec = sumdminus/(totalobs - totaldisease);
run;
```

In the preceding code, `proc summary` (section 2.1.1) is used to find the total number with the disease and in the dataset, and to save this data in a dataset named `sumdisease`. The data is then sorted in descending order of the test score `t`. In the final step, the disease and total number of observations are read in and the current number of true positives and negatives accrued as the value of `t` decreases. The conditional use of the `set` statement allows the summary values for disease and subjects to be included for each line of the output dataset; the `retain` statement allows values to be kept across entries in the dataset and optionally allows the initial value to be set. The final dataset contains the sensitivity `sens` and 1 minus the specificity `one_m_spec`. This approach would be more complicated if tied values of the test score were possible.

R

```
sens <- sum(D==1&T==1)/sum(D==1)
spec <- sum(D==0&T==0)/sum(D==0)
```

Note: Sensitivity and specificity for an outcome `D` can be calculated for each value of a continuous measure `T` using the following code.

```
library(ROCR)
pred <- prediction(T, D)
diagobj <- performance(pred, "sens", "spec")
spec <- slot(diagobj, "y.values")[[1]]
sens <- slot(diagobj, "x.values")[[1]]
cut <- slot(diagobj, "alpha.values")[[1]]
diagmat <- cbind(cut, sens, spec)
head(diagmat, 10)
```

Note: The `ROCR` package facilitates the calculation of test characteristics, including sensitivity and specificity. The `prediction()` function takes as arguments the continuous measure and outcome. The returned object can be used to calculate quantities of interest (see `help(performance)` for a comprehensive list). The `slot()` function is used to return the desired sensitivity and specificity values for each cut score, where `[[1]]` denotes the first element of the returned list (see `help(list)` and `help(Extract)`).

2.2.3 Correlation

HELP example: see 2.6.2 and 5.6.6

SAS

```
proc corr data=ds;
   var x1 ... xk;
run;
```

Note: Specifying `spearman` or `kendall` as an option to `proc corr` generates the Spearman or Kendall correlation coefficients, respectively. The `with` statement can be used to generate correlations only between the `var` and `with` variables, as in 2.6.2, rather than among all the `var` variables. This can save space as it avoids replicating correlations above and below the diagonal of the correlation matrix.

R

```
pearsoncorr <- cor(x, y)
spearmancorr <- cor(x, y, method="spearman")
kendalltau <- cor(x, y, method="kendall")
```

or

```
cormat <- cor(cbind(x1, ..., xk))
```

Note: Specifying `method="spearman"` or `method="kendall"` as an option to `cor()` generates the Spearman or Kendall correlation coefficients, respectively. A matrix of variables (created with `cbind()`) can be used to generate the correlation between a set of variables. To emulate the `with` statement in SAS, subsets of the returned correlation matrix can be selected, as demonstrated in section 2.6.2. This can save space as it avoids replicating correlations above and below the diagonal of the correlation matrix. The `use` option for `cor()` specifies how missing values are handled (either `"all.obs"`, `"complete.obs"` or `"pairwise.complete.obs"`).

2.2.4 Kappa (agreement)

SAS

```
proc freq data=ds;
   tables x * y / agree;
run;
```

Note: The `agree` statement produces κ and weighted κ and their asymptotic standard errors and confidence interval, as well as McNemar's test for 2×2 tables, and Bowker's test of symmetry for tables with more than two levels [7].

R

```
library(irr)
kappa2(data.frame(x, y))
```

Note: The `kappa2()` function takes a dataframe (see B.4.5) as argument. Weights can be specified as an option.

2.3 Contingency tables

2.3.1 Display cross-classification table

HELP example: see 2.6.3

Contingency tables show the group membership across categorical (grouping) variables. They are also known as cross-classification tables, cross-tabulations, and two-way tables.

SAS

```
proc freq data=ds;
   tables x * y;
run;
```

R

```
mytab <- table(y, x)
addmargins(mytab)
prop.table(mytab, 1)
```

or

```
xtabs(~ y + x)
```

or

```
library(prettyR)
xtab(y ~ x, data=ds)
```

Note: The `addmargins()` function adds (by default) the row and column totals to a table, while `prop.table()` can be used to calculate row totals (with option 1) and column totals (with option 2). The `colSums()`, `colMeans()` functions (and their equivalents for rows) can be used to efficiently calculate sums and means for numeric vectors. The `xtabs()` function can be used to create a contingency table from cross-classifying factors. Much of the process of displaying tables is automated in the `prettyR` library `xtab()` function.

2.3.2 Pearson chi-square statistic

HELP example: see 2.6.3

SAS

```
proc freq data=ds;
   tables x * y / chisq;
run;
```

Note: For 2×2 tables the output includes both unadjusted and continuity-corrected tests.

R

```
chisq.test(x, y)
```

Note: The `chisq.test()` command can accept either two class vectors or a matrix with counts. By default a continuity correction is used (the option `correct=FALSE` turns this off).

2.3.3 Cochran–Mantel–Haenszel test

The Cochran–Mantel–Haenszel test gives an assessment of the relationship between X_2 and X_3, stratified by (or controlling for) X_1. The analysis provides a way to adjust for the possible confounding effects of X_1 without having to estimate parameters for them.

SAS

```
proc freq data=ds;
   tables x1 * x2 * x3 / cmh;
run;
```

Note: The `cmh` option produces Cochran–Mantel–Haenszel statistics and, when both X_2 and X_3 have two values, it generates estimates of the common odds ratio, common relative risks, and the Breslow–Day test for homogeneity of the odds ratios. More complex models can be fit using the generalized linear model methodology described in Chapter 4.

R

```
mantelhaen.test(x2, x3, x1)
```

2.3.4 Fisher's exact test

SAS *HELP example:* see 2.6.3

```
proc freq data=ds;
   tables x * y / exact;
run;
```

or

```
proc freq data=ds;
   tables x * y;
   exact fisher / mc n=bnum;
run;
```

Note: The former requests only the exact p-value; the latter generates a Monte Carlo p-value, an asymptotically equivalent test based on `bnum` random tables simulated using the observed margins.

R

```
fisher.test(y, x)
```

or

```
fisher.test(ymat)
```

Note: The `fisher.test()` command can accept either two class vectors or a matrix with counts (here denoted by `ymat`). For tables with many rows and/or columns, p-values can be computed using Monte Carlo simulation using the `simulate.p.value` option.

2.3.5 McNemar's test

McNemar's test tests the null hypothesis that the proportions are equal across matched pairs, for example, when two raters assess a population.

SAS

```
proc freq data=ds;
   tables x * y / agree;
run;
```

R

```
mcnemar.test(y, x)
```

Note: The `mcnemar.test()` command can accept either two class vectors or a matrix with counts.

2.4 Two sample tests for continuous variables

2.4.1 Student's t-test

SAS *HELP example:* see 2.6.4

```
proc ttest data=ds;
   class x;
   var y;
run;
```

Note: The variable X takes on two values. The output contains both equal and unequal-variance t-tests, as well as a test of the null hypothesis of equal variance.

R

```
t.test(y1, y2)
```

or

```
t.test(y ~ x)
```

Note: The first example for the `t.test()` command displays how it can take two vectors (`y1` and `y2`) as arguments to compare, or in the latter example a single vector corresponding to the outcome (`y`), with another vector indicating group membership (`x`) using a formula interface (see sections B.4.6 and 3.1.1). By default, the two-sample t-test uses an unequal variance assumption. The option `var.equal=TRUE` can be added to specify an equal variance assumption. The command `var.test()` can be used to formally test equality of variances.

2.4.2 Nonparametric tests

HELP example: see 2.6.4

SAS

```
proc npar1way data=ds wilcoxon edf median;
   class y;
   var x;
run;
```

Note: Many tests can be requested as options to the `proc npar1way` statement. Here we show a Wilcoxon test, a Kolmogorov–Smirnov test, and a median test, respectively. Exact tests can be generated by using an `exact` statement with these names, e.g., the `exact median` statement will generate the exact median test.

R

```
wilcox.test(y1, y2)
ks.test(y1, y2)

library(coin)
median_test(y ~ x)
```

Note: By default, the `wilcox.test()` function uses a continuity correction in the normal approximation for the p-value. The `ks.test()` function does not calculate an exact p-value when there are ties. The median test shown will generate an exact p-value with the `distribution="exact"` option.

2.4.3 Permutation test

HELP example: see 2.6.4

SAS

```
proc npar1way data=ds;
   class y;
   var x;
   exact scores=data;
run;
```

or

```
proc npar1way data=ds;
   class y;
   var x;
   exact scores=data / mc n=bnum;
run;
```

Note: Any test described in 2.4.2 can be named in place of `scores=data` to get an exact test based on those statistics. The `mc` option generates an empirical p-value (asymptotically equivalent to the exact p-value) based on `bnum` Monte Carlo replicates.

R

```
library(coin)
oneway_test(y ~ as.factor(x), distribution=approximate(B=bnum))
```

Note: The `oneway_test` function in the `coin` library implements a variety of permutation based tests (see also the `exactRankTests` package). The `distribution=approximate` syntax generates an empirical p-value (asymptotically equivalent to the exact p-value) based on `bnum` Monte Carlo replicates.

2.4.4 Logrank test

HELP example: see 2.6.5

See also 5.1.19 (Kaplan–Meier plot) and 4.3.1 (Cox proportional hazards model)

SAS

```
proc phreg data=ds;
   model timevar*cens(0) = x;
run;
```

or

```
proc lifetest data=ds;
   time timevar*cens(0);
   strata x;
run;
```

Note: If `cens` is equal to 0, then `proc phreg` and `proc lifetest` treat *time* as the time of censoring, otherwise it is the time of the event. The default output from `proc lifetest` includes the logrank and Wilcoxon tests. Other tests, corresponding to different weight functions, can be produced with the `test` option to the `strata` statement. These include `test=fleming(`ρ_1, ρ_2`)`, a superset of the G-rho family of Fleming and Harrington [23], which simplifies to the G-rho family when $\rho_2 = 0$.

R

```
library(survival)
survdiff(Surv(timevar, cens) ~ x)
```

Note: Other tests within the G-rho family of Fleming and Harrington [23] are supported by specifying the `rho` option.

2.5 Further resources

Comprehensive introductions to using SAS to fit common statistical models can be found in [9] and [15]. Similar methods in R are accessibly presented in [95]. Efron and Tibshirani [18] provides a comprehensive overview of bootstrapping. A readable introduction to permutation-based inference can be found in [27]. Collett [11] is an accessible introduction to survival analysis.

2.6 HELP examples

To help illustrate the tools presented in this chapter, we apply many of the entries to the HELP data. SAS and R code can be downloaded from `http://www.math.smith.edu/sasr/examples`.

2.6.1 Summary statistics and exploratory data analysis

We begin by reading the dataset.

```
filename myurl
   url 'http://www.math.smith.edu/sasr/datasets/help.csv'  lrecl=704;

proc import
   datafile=myurl
   out=ds dbms=dlm;
   delimiter=',';
   getnames=yes;
run;
```

The `lrecl` statement is needed due to the long lines in the csv file.

```
> options(digits=3)
> options(width=72)  # narrows output to stay in the grey box
> ds <- read.csv("http://www.math.smith.edu/sasr/datasets/help.csv")
> attach(ds)
```

A first step would be to examine some univariate statistics (2.1.1) for the baseline CESD (Center for Epidemiologic Statistics measure of depressive symptoms) score. In SAS, univariate statistics are produced by `proc univariate`, `proc means`, and others.

```
options ls=70;  * narrow output to stay in grey box;
proc means data=ds maxdec=2 min p5 q1 median q3 p95 max mean std range;
   var cesd;
run;

The MEANS Procedure

                      Analysis Variable : cesd

                                  Lower                        Upper
    Minimum        5th Pctl    Quartile          Median     Quartile
---------------------------------------------------------------------
       1.00           10.00       25.00           34.00        41.00
---------------------------------------------------------------------

                      Analysis Variable : cesd

  95th Pctl         Maximum          Mean         Std Dev        Range
---------------------------------------------------------------------
      53.00           60.00         32.85           12.51        59.00
---------------------------------------------------------------------
```

In R, we can use functions which produce a set of statistics, such as `fivenum()`, or request them singly.

```
> fivenum(cesd)
[1]  1 25 34 41 60
> mean(cesd); median(cesd)
[1] 32.8
[1] 34
> range(cesd)
[1]  1 60
> sd(cesd)
[1] 12.5
> var(cesd)
[1] 157
```

We can also generate desired statistics. Here, we find the deciles (2.1.5).

```
ods select none;
proc univariate data=ds;
   var cesd;
   output out=deciles pctlpts= 0 to 100 by 10 pctlpre=p_;
run;
ods select all;

options ls=74;
proc print data=deciles;
run;

Obs  p_0  p_10  p_20  p_30  p_40  p_50  p_60  p_70  p_80  p_90  p_100

 1    1    15    22    27    30    34    37    40    44    49     60
```

```
> quantile(cesd, seq(from=0, to=1, length=11))
  0%  10%  20%  30%  40%  50%  60%  70%  80%  90% 100%
 1.0 15.2 22.0 27.0 30.0 34.0 37.0 40.0 44.0 49.0 60.0
```

Graphics can allow us to easily review the whole distribution of the data. Here we generate a histogram (5.1.4) of CESD, overlaid with its empirical PDF (5.1.16) and the closest-fitting normal distribution (see Figure 2.1). In SAS, the other results of `proc univariate` have been suppressed by `selecting` only the graphics output using an `ods select` statement (note the different y-axes generated).

```
ods select univar;
proc univariate data=ds;
   var cesd;
   histogram cesd / normal (color=black l=1) kernel(color=black l=21)
                     cfill=greyCC;
run; quit;
ods select all;
```

```
> hist(cesd, main="", freq=FALSE)
> lines(density(cesd), main="CESD", lty=2, lwd=2)
> xvals <- seq(from=min(cesd), to=max(cesd), length=100)
> lines(xvals, dnorm(xvals, mean(cesd), sd(cesd)), lwd=2)
```

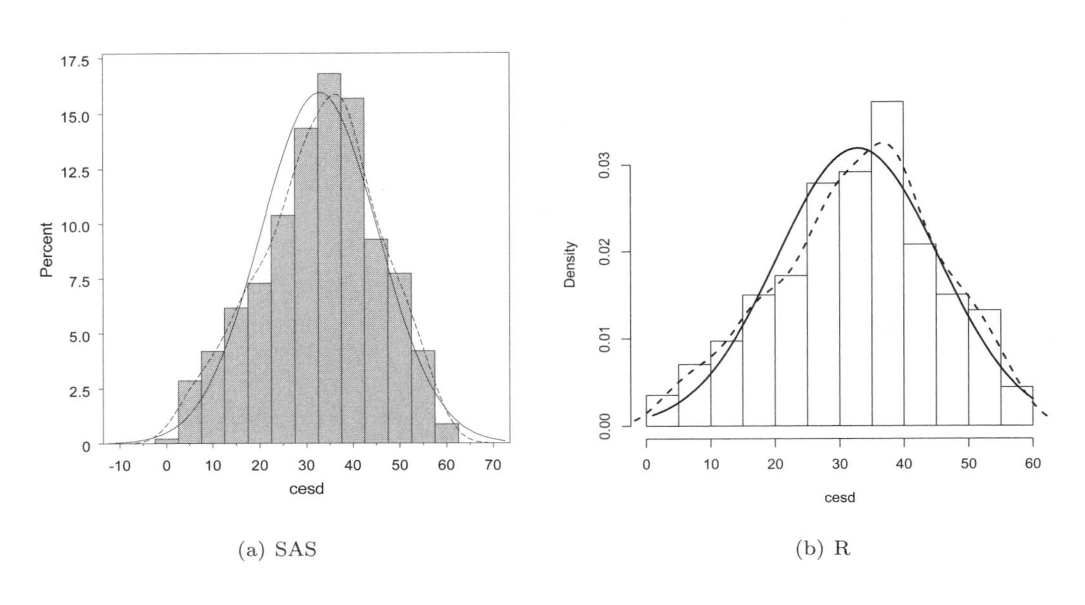

(a) SAS (b) R

Figure 2.1: Density plot of depressive symptom scores (CESD) plus superimposed histogram and normal distribution

2.6.2 Bivariate relationships

We can calculate the correlation (2.2.3) between CESD and MCS and PCS (mental and physical component scores). First, we show the default correlation matrix.

```
ods select pearsoncorr;
proc corr data=ds;
   var cesd mcs pcs;
run;

The CORR Procedure

                 cesd             mcs             pcs

cesd          1.00000        -0.68192        -0.29270
                               <.0001          <.0001

mcs          -0.68192         1.00000         0.11046
               <.0001                          0.0187

pcs          -0.29270         0.11046         1.00000
               <.0001          0.0187
```

```
> cormat <- cor(cbind(cesd, mcs, pcs))
> cormat

       cesd    mcs    pcs
cesd  1.000 -0.682 -0.293
mcs  -0.682  1.000  0.110
pcs  -0.293  0.110  1.000
```

To save space, we can just print a subset of the correlations.

```
ods select pearsoncorr;
proc corr data=ds;
   var mcs pcs;
   with cesd;
run;

The CORR Procedure

                  mcs             pcs

cesd         -0.68192        -0.29270
               <.0001          <.0001
```

```
> cormat[c(2, 3), 1]

   mcs    pcs
-0.682 -0.293
```

Figure 2.2 displays a scatterplot (5.1.1) of CESD and MCS, for the female subjects. The plotting character (5.2.2) is the primary substance (Alcohol, Cocaine, or Heroin). For R, a rug plot (5.2.8) is added to help demonstrate the marginal distributions; this is nontrivial in SAS.

```
symbol1 font=swiss v='A' h=.7 c=black;
symbol2 font=swiss v='C' h=.7 c=black;
symbol3 font=swiss v='H' h=.7 c=black;
proc gplot data=ds;
   where female=1;
   plot mcs*cesd=substance;
run; quit;
```

```
> plot(cesd[female==1], mcs[female==1], xlab="CESD", ylab="MCS",
+     type="n", bty="n")
> text(cesd[female==1&substance=="alcohol"],
+     mcs[female==1&substance=="alcohol"],"A")
> text(cesd[female==1&substance=="cocaine"],
+     mcs[female==1&substance=="cocaine"],"C")
> text(cesd[female==1&substance=="heroin"],
+     mcs[female==1&substance=="heroin"],"H")
> rug(jitter(mcs[female==1]), side=2)
> rug(jitter(cesd[female==1]), side=3)
```

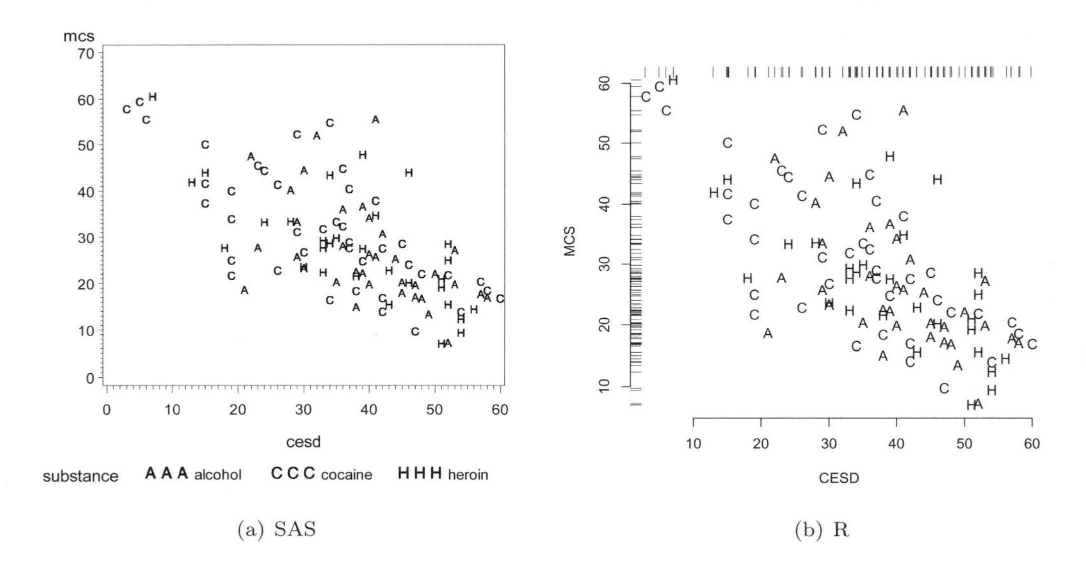

(a) SAS (b) R

Figure 2.2: Scatterplot of CESD and MCS for women, with primary substance shown as the plot symbol

2.6.3 Contingency tables

Here we display the cross-classification (contingency) table (2.3.1) of homeless at baseline by gender, calculate the observed odds ratio (OR, 2.2.1), and assess association using the Pearson χ^2 test (2.3.2) and Fisher's exact test (2.3.4). In SAS, this can be done with one call to `proc freq`.

```
proc freq data=ds;
   tables homeless*female / chisq exact relrisk;
run; quit;
```

```
The FREQ Procedure

Table of homeless by female

homeless     female

Frequency|
Percent  |
Row Pct  |
Col Pct  |       0|       1|  Total
---------+--------+--------+
       0 |    177 |     67 |    244
         |  39.07 |  14.79 |  53.86
         |  72.54 |  27.46 |
         |  51.16 |  62.62 |
---------+--------+--------+
       1 |    169 |     40 |    209
         |  37.31 |   8.83 |  46.14
         |  80.86 |  19.14 |
         |  48.84 |  37.38 |
---------+--------+--------+
Total         346      107      453
            76.38    23.62   100.00
```

```
Statistics for Table of homeless by female

Statistic                     DF       Value      Prob
------------------------------------------------------
Chi-Square                     1      4.3196    0.0377
Likelihood Ratio Chi-Square    1      4.3654    0.0367
Continuity Adj. Chi-Square     1      3.8708    0.0491
Mantel-Haenszel Chi-Square     1      4.3101    0.0379
Phi Coefficient                      -0.0977
Contingency Coefficient               0.0972
Cramer's V                           -0.0977
```

Fisher's exact test is provided by default with 2×2 tables, so the `exact` statement is not required. The exact test result is shown.

```
Statistics for Table of homeless by female

Cell (1,1) Frequency (F)       177
Left-sided Pr <= F          0.0242
Right-sided Pr >= F         0.9861

Table Probability (P)       0.0102
Two-sided Pr <= P           0.0456
```

```
Statistics for Table of homeless by female

           Estimates of the Relative Risk (Row1/Row2)

Type of Study                     Value       95% Confidence Limits
-----------------------------------------------------------------
Case-Control (Odds Ratio)        0.6253        0.4008        0.9755
Cohort (Col1 Risk)               0.8971        0.8105        0.9930
Cohort (Col2 Risk)               1.4347        1.0158        2.0265
```

In R, the `table()` function can display contingency tables. The `prettyR` library provides a way to display them with additional statistics, similar to SAS.

```
> table(homeless, female)

        female
homeless   0   1
       0 177  67
       1 169  40
> library(prettyR)
> xtres <- xtab(homeless ~ female, data=ds)

Crosstabulation of homeless by female
         female
homeless          0         1
0               177        67       244
              72.54     27.46     53.86
              51.16     62.62

1               169        40       209
              80.86     19.14     46.14
              48.84     37.38

                346       107       453
              76.38     23.62
```

In R, we can easily calculate the odds ratio directly. If the odds ratio were not available from a procedure in SAS, it would require several steps to replicate these calculations.

```
> or <- (sum(homeless==0 & female==0)*
+        sum(homeless==1 & female==1))/
+       (sum(homeless==0 & female==1)*
+        sum(homeless==1 & female==0))
> or

[1] 0.625
```

```
> library(epitools)
> oddsobject <- oddsratio.wald(homeless, female)
> oddsobject$measure

         odds ratio with 95% C.I.
Predictor estimate lower upper
        0    1.000    NA    NA
        1    0.625 0.401 0.975

> oddsobject$p.value

         two-sided
Predictor midp.exact fisher.exact chi.square
        0         NA           NA         NA
        1     0.0381       0.0456     0.0377
```

The χ^2 and Fisher's exact tests are fit in R using separate commands.

```
> chisqval <- chisq.test(homeless, female, correct=FALSE)
> chisqval

        Pearson's Chi-squared test

data:  homeless and female
X-squared = 4.32, df = 1, p-value = 0.03767
```

```
> fisher.test(homeless, female)

        Fisher's Exact Test for Count Data

data:  homeless and female
p-value = 0.04560
alternative hypothesis: true odds ratio is not equal to 1
95 percent confidence interval:
 0.389 0.997
sample estimates:
odds ratio
     0.626
```

2.6.4 Two sample tests of continuous variables

We can assess gender differences in baseline age using a t-test (2.4.1) and nonparametric procedures.

```
options ls=74;  /* narrows output to stay in the grey box */

proc ttest data=ds;
   class female;
   var age;
run;

Variable:  age

female           N        Mean     Std Dev     Std Err     Minimum     Maximum

0              346     35.4682      7.7501      0.4166     19.0000     60.0000
1              107     36.2523      7.5849      0.7333     21.0000     58.0000
Diff (1-2)             -0.7841      7.7116      0.8530

female        Method                Mean       95% CL Mean          Std Dev

0                                35.4682     34.6487  36.2877        7.7501
1                                36.2523     34.7986  37.7061        7.5849
Diff (1-2)    Pooled             -0.7841     -2.4605   0.8923        7.7116
Diff (1-2)    Satterthwaite      -0.7841     -2.4483   0.8800

female        Method             95% CL Std Dev

0                                7.2125   8.3750
1                                6.6868   8.7637
Diff (1-2)    Pooled             7.2395   8.2500
Diff (1-2)    Satterthwaite
Method           Variances        DF    t Value    Pr > |t|

Pooled           Equal           451      -0.92      0.3585
Satterthwaite    Unequal      179.74      -0.93      0.3537

             Equality of Variances

Method      Num DF    Den DF    F Value    Pr > F

Folded F       345       106       1.04    0.8062
> ttres <- t.test(age ~ female, data=ds)
> print(ttres)

        Welch Two Sample t-test

data:  age by female
t = -0.93, df = 180, p-value = 0.3537
alternative hypothesis: true difference in means is not equal to 0
95 percent confidence interval:
 -2.45  0.88
sample estimates:
mean in group 0 mean in group 1
           35.5            36.3
```

The `names()` function can be used to identify the objects returned by the `t.test()` function (not displayed).

A permutation test can be run and used to generate a Monte Carlo p-value (2.4.3).

```
ods select datascoresmc;
proc npar1way data=ds;
   class female;
   var age;
   exact scores=data / mc n=9999 alpha=.05;
run;
ods exclude none;

One-Sided Pr >= S
Estimate                          0.1789
95% Lower Conf Limit              0.1714
95% Upper Conf Limit              0.1864

Two-Sided Pr >= |S - Mean|
Estimate                          0.3557
95% Lower Conf Limit              0.3464
95% Upper Conf Limit              0.3651

Number of Samples                   9999
Initial Seed                   998734001
```

```
> library(coin)
> oneway_test(age ~ as.factor(female),
+    distribution=approximate(B=9999), data=ds)

        Approximative 2-Sample Permutation Test

data:  age by as.factor(female) (0, 1)
Z = -0.92, p-value = 0.3592
alternative hypothesis: true mu is not equal to 0
```

Both the Wilcoxon test and Kolmogorov–Smirnov test (2.4.2) can be run with a single call to `proc freq`. Later, we'll include the D statistic from the Kolmogorov–Smirnov test and the associated p-value in a Figure title; to make that possible, we'll use ODS to create a dataset containing these values.

```
ods output kolsmir2stats=age_female_ks_stats;
ods select wilcoxontest kolsmir2stats;
proc npar1way data=ds wilcoxon edf;
   class female;
   var age;
run;
ods select all;
```

```
Statistic                25288.5000

Normal Approximation
Z                              0.8449
One-Sided Pr >  Z              0.1991
Two-Sided Pr > |Z|             0.3981

t Approximation
One-Sided Pr >  Z              0.1993
Two-Sided Pr > |Z|             0.3986

Z includes a continuity correction of 0.5.
```

```
KS    0.026755     D           0.062990
KSa   0.569442     Pr > KSa    0.9020
```

In R, these tests are obtained in separate function calls (see 2.4.2).

```
> wilcox.test(age ~ as.factor(female), correct=FALSE)

        Wilcoxon rank sum test

data:  age by as.factor(female)
W = 17512, p-value = 0.3979
alternative hypothesis: true location shift is not equal to 0
```

```
> ksres <- ks.test(age[female==1], age[female==0], data=ds)
> print(ksres)

        Two-sample Kolmogorov-Smirnov test

data:  age[female == 1] and age[female == 0]
D = 0.063, p-value = 0.902
alternative hypothesis: two-sided
```

We can also plot estimated density functions (5.1.16) for age for both groups, and shade some areas (5.2.13) to emphasize how they overlap (Figure 2.3). SAS `proc univariate` with a `by` statement will generate density estimates for each group, but not over-plot them. To get results similar to those available through R, we first generate the density estimates using `proc kde` (5.1.16) (suppressing all printed output).

```
proc sort data=ds;
   by female;
run;

ods select none;
proc kde data=ds;
   by female;
   univar age / out=kdeout;
run;
ods select all;
```

Next, we'll review the `proc npar1way` output which was saved as a dataset.

```
proc print data=age_female_ks_stats; run;

   V
   a                 c                         n                    c           n
   r           L     V                         V         L          V           V
   i     N     a     a                         a    N    a          a           a
   a     a     b     l                         l    a    b          l           l
O  b     m     e     u                         u    m    e          u           u
b  l     e     l     e                         e    e    l          e           e
s  e     1     1     1                         1    2    2          2           2

1 age _KS_  KS  0.026755        0.026755 _D_   D          0.062990     0.062990
2 age _KSA_ KSa 0.569442        0.569442 P_KSA Pr > KSa 0.9020       0.901979
```

Running `proc contents` (1.3.2, results not shown) reveals that the variable names prepended with 'c' are character variables. To get these values into a Figure title, we use SAS Macro variables (A.8.2) created by the `call symput` function.

```
data _null_;
set age_female_ks_stats;
   if label2 eq 'D' then call symput('dvalue', substr(cvalue2, 1, 5));
     /* This makes a macro variable (which is saved outside any dataset)
        from a value in a dataset */
   if label2 eq 'Pr > KSa' then call symput('pvalue', substr(cvalue2, 1, 4));
run;
```

Finally, we construct the plot using `proc gplot` for the data with a `title` statement to include the Kolmogorov–Smirnov test results.

```
symbol1 i=j w=5 l=1 v=none c=black;
symbol2 i=j w=5 l=2 v=none c=black;
title "Test of ages: D=&dvalue p=&pvalue";
pattern1 color=grayBB;
proc gplot data=kdeout;
   plot density*value = female / legend areas=1 haxis=18 to 60 by 2;
run; quit;
```

In this code, the `areas` option to the `plot` statement makes SAS fill in the area under the first curve, while the `pattern` statement describes what color to fill in with.

In R, we can create a function (see B.5) to automate this task.

```
> plotdens <- function(x,y, mytitle, mylab) {
+    densx <- density(x)
+    densy <- density(y)
+    plot(densx, main=mytitle, lwd=3, xlab=mylab, bty="l")
+    lines(densy, lty=2, col=2, lwd=3)
+    xvals <- c(densx$x, rev(densy$x))
+    yvals <- c(densx$y, rev(densy$y))
+    polygon(xvals, yvals, col="gray")
+ }
```

The `polygon()` function is used to fill in the area between the two curves.

```
> mytitle <- paste("Test of ages: D=", round(ksres$statistic, 3),
+    " p=", round(ksres$p.value, 2), sep="")
> plotdens(age[female==1], age[female==0], mytitle=mytitle,
+    mylab="age (in years)")
> legend(50, .05, legend=c("Women", "Men"), col=1:2, lty=1:2, lwd=2)
```

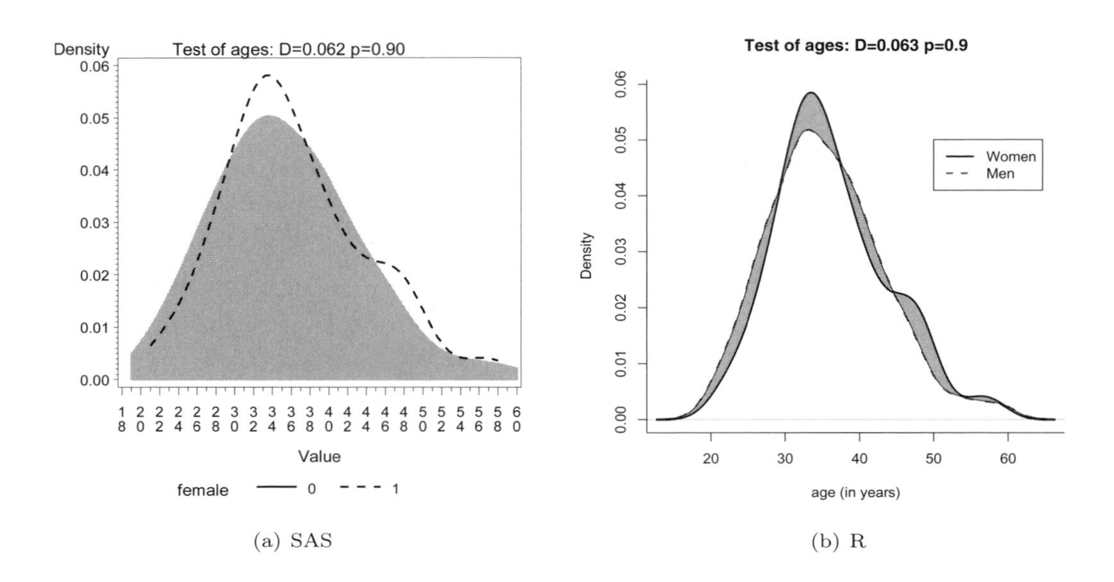

(a) SAS (b) R

Figure 2.3: Density plot of age by gender

2.6.5 Survival analysis: logrank test

The logrank test (2.4.4) can be used to compare estimated survival curves between groups in the presence of censoring. Here we compare randomization groups with respect to `dayslink`, where a value of 0 for `linkstatus` indicates that the observation was censored, not observed, at the time recorded in `dayslink`.

```
ods select homtests;
proc lifetest data=ds;
   time dayslink*linkstatus(0);
   strata treat;
run;
ods select all;

         Test of Equality over Strata

                                          Pr >
Test         Chi-Square       DF     Chi-Square

Log-Rank        84.7878        1         <.0001
Wilcoxon        87.0714        1         <.0001
-2Log(LR)      107.2920        1         <.0001
```

```
> library(survival)
> survobj <- survdiff(Surv(dayslink, linkstatus) ~ treat,
+    data=ds)
> print(survobj)

Call:
survdiff(formula = Surv(dayslink, linkstatus) ~ treat, data = ds)

n=431, 22 observations deleted due to missingness.

          N Observed Expected (O-E)^2/E (O-E)^2/V
treat=0 209       35     92.8      36.0      84.8
treat=1 222      128     70.2      47.6      84.8

 Chisq= 84.8  on 1 degrees of freedom, p= 0

> names(survobj)

[1] "n"         "obs"       "exp"       "var"       "chisq"
[6] "na.action" "call"
```

Chapter 3

Linear regression and ANOVA

Regression and analysis of variance form the basis of many investigations. In this chapter we describe how to undertake many common tasks in linear regression (broadly defined), while chapter 4 discusses many generalizations, including other types of outcome variables, longitudinal and clustered analysis, and survival methods.

Many SAS procedures and R commands can perform linear regression, as it constitutes a special case of which many models are generalizations. We present detailed descriptions for SAS `proc reg` and `proc glm` as well as for the R `lm()` command, as these offer the most flexibility and best output options tailored to linear regression in particular. While ANOVA can be viewed as a special case of linear regression, separate routines are available in SAS (`proc anova`) and R (`aov()`) to perform it. In addition, SAS `proc mixed` is also useful for some calculations. We address these additional procedures only with respect to output that is difficult to obtain through the standard linear regression tools.

Many of the routines available within R return or operate on `lm` class objects, which includes objects such as coefficients, residuals, fitted values, weights, contrasts, model matrices, and the like (see `help(lm)`).

The CRAN Task View on Statistics for the Social Sciences provides an excellent overview of methods described here and in Chapter 4.

3.1 Model fitting

3.1.1 Linear regression

SAS

HELP example: see 3.7.2

```
proc glm data=ds;
   model y = x1 ... xk;
run;
```

or

```
proc reg data=ds;
   model y = x1 ... xk;
run;
```

Note: Both `proc glm` and `proc reg` support linear regression models, while `proc reg` provides more regression diagnostics. The `glm` procedure more easily allows categorical covariates.

R

```
mod1 <- lm(y ~ x1 + ... + xk, data=ds)
summary(mod1)
```

or

```
form <- as.formula(y ~ x1 + ... + xk)
mod1 <- lm(form, data=ds)
summary(mod1)
```

Note: The first argument of the `lm()` function is a formula object, with the outcome specified followed by the ~ operator then the predictors. More information about the linear model `summary()` command can be found using `help(summary.lm)`. By default, stars are used to annotate the output of the `summary()` functions regarding significance levels: these can be turned off using the command `options(show.signif.stars=FALSE)`.

3.1.2 Linear regression with categorical covariates

HELP example: see 3.7.2

See also 3.1.3 (parameterization of categorical covariates)

SAS

```
proc glm data=ds;
   class x1;
   model y = x1 x2 ... xk;
run;
```

Note: The `class` statement specifies covariates that should be treated as categorical. The `glm` procedure uses reference cell coding; the reference category can be controlled using the `order` option to the `proc glm` statement, as in 4.6.11.

R

```
x1f <- as.factor(x1)
mod1 <- lm(y ~ x1f + x2 + ... + xk, data=ds)
```

Note: The `as.factor()` command in R creates a categorical variable from a variable. By default, the lowest value (either numerically or lexicographically) is the reference value. The `levels` option for the `factor()` function can be used to select a particular reference value (see also 1.4.12).

3.1.3 Parameterization of categorical covariates

HELP example: see 3.7.5

SAS and R handle this issue in different ways. In R, `as.factor()` can be applied within any model-fitting function. Parameterization of the covariate can be controlled as below. For SAS, some procedures accept a `class` statement to declare that a covariate to be treated as categorical. As of SAS version 9.2, of the model-fitting procedures mentioned in this book, the following procedures will not accept a class statement: `arima`, `catmod`, `countreg`, `factor`, `freq`, `kde`, `lifetest`, `nlin`, `nlmixed`, `reg`, `surveyfreq`, and `varclus`. For these procedures, indicator (or "dummy") variables must be created in a data step. The following procedures accept a class statement which applies reference cell or indicator variable coding (described as `contr.SAS()` in the R note below) to the listed variables: `proc anova`, `candisc`, `discrim`, `gam`, `glimmix`, `glm`, `mi`, `mianalyze`, `mixed`, `quantreg`, `robustreg`, `stepdisc`, and `surveyreg`. The value used as the referent can often be controlled, usually as an `order` option to the controlling `proc`, as in 4.6.11. For these procedures, other parameterizations must be coded in a data step. The following procedures accept multiple parameterizations, using the syntax shown below for `proc logistic`: `proc genmod` (defaults to reference cell coding), `proc logistic` (defaults to effect coding), `proc phreg` (defaults to reference cell coding), and `proc surveylogistic` (defaults to effect coding).

SAS

```
proc logistic data=ds;
   class x1 (param=paramtype) x2 (param=paramtype);
   ...
run;
```

or

```
proc logistic data=ds;
   class x1 x2 / param=paramtype;
   ...
run;
```

Note: Available `paramtypes` include: 1) `orthpoly`, equivalent to `contr.poly()`; 2) `effect` (the default for `procs logistic` and `surveylogistic`), equivalent to `contr.sum()`; and 3) `ref`, equivalent to `contr.SAS()`. In addition, if the same parameterization is desired for all of the categorical variables in the model, it can be added in a statement such as the second example. In this case, `param=glm` can be used to emulate the parameterization found in the other procedures which accept `class` statements and in `contr.SAS()` within R; this is (the default for `procs genmod` and `phreg`).

R

```
x.factor <- as.factor(x)
mod1 <- lm(y ~ x.factor, contrasts=list(x.factor="contr.SAS"))
```

Note: The `as.factor()` function creates a factor object, akin to how SAS treats class variables in `proc glm`. The `contrasts` option for the `lm()` function specifies how the levels of that factor object should be used within the function. The `levels` option to the `factor()` function allows specification of the ordering of levels (the default is lexicographic). An example can be found in section 3.7.

The specification of the design matrix for analysis of variance and regression models can be controlled using the `contrasts` option. Examples of options (for a factor with 4 equally-spaced levels) are given below.

```
> contr.treatment(4)
  2 3 4
1 0 0 0
2 1 0 0
3 0 1 0
4 0 0 1
> contr.SAS(4)
  1 2 3
1 1 0 0
2 0 1 0
3 0 0 1
4 0 0 0
```

```
> contr.poly(4)
         .L   .Q     .C
[1,] -0.671  0.5 -0.224
[2,] -0.224 -0.5  0.671
[3,]  0.224 -0.5 -0.671
[4,]  0.671  0.5  0.224
> contr.sum(4)
  [,1] [,2] [,3]
1    1    0    0
2    0    1    0
3    0    0    1
4   -1   -1   -1
```

```
> contr.helmert(4)
  [,1] [,2] [,3]
1   -1   -1   -1
2    1   -1   -1
3    0    2   -1
4    0    0    3
```

See `options("contrasts")` for defaults, and `contrasts()` or `lm()` to apply a contrast function to a factor variable. Support for reordering factors is available within the `reshape` library `reorder_factor()` function.

3.1.4 Linear regression with no intercept

SAS

```
proc glm data=ds;
   model y = x1 ... xk / noint;
run;
```

Note: The `noint` option works with many `model` statements.

R

```
mod1 <- lm(y ~ 0 + x1 + ... + xk, data=ds)
```

or

```
mod1 <- lm(y ~ x1 + ... + xk -1, data=ds)
```

3.1.5 Linear regression with interactions

HELP example: see 3.7.2

SAS

```
proc glm data=ds;
   model y = x1 x2 x1*x2 x3 ... xk;
run;
```

or

```
proc glm data=ds;
   model y = x1|x2 x3 ... xk;
run;
```

Note: The `|` operator includes the product and all lower order terms, while the `*` operator includes only the specified interaction. So, for example, `model y = x1|x2|x3` and `model y = x1 x2 x3 x1*x2 x1*x3 x2*x3 x1*x2*x3` are equivalent statements. The syntax above also works with any covariates designated as categorical using the `class` statement (3.1.2). The `model` statement for many procedures accepts this syntax.

R

```
mod1 <- lm(y ~ x1 + x2 + x1:x2 + x3 + ... + xk, data=ds)
```

or

```
lm(y ~ x1*x2 + x3 + ... + xk, data=ds)
```

Note: The `*` operator includes all lower order terms (in this case main effects), while the `:` operator includes only the specified interaction. So, for example, the commands `y ~ x1*x2*x3` and `y ~ x1 + x2 + x3 + x1:x2 + x1:x3 + x2:x3 + x1:x2:x3` have equal values. The syntax also works with any covariates designated as categorical using the `as.factor()` command (see 3.1.2).

3.1.6 Linear models stratified by each value of a grouping variable

HELP example: see 3.7.4

It is straightforward in SAS to fit models stratified by each value of a grouping variable. In R this task is more complicated (see also subsetting, 1.5.1).

SAS

```
proc sort data=ds;
   by z;
run;

ods output parameterestimates=params;
proc reg data=ds;
   by z;
   model y = x1 ... xk;
run;
```

Note: Note that if the `by` variable has many distinct values, output may be voluminous. A single dataset containing the parameter estimates from each `by` group (A.6.2) can be created using `ODS` by issuing an `ods output parameterestimates=ds` statement before the `proc reg` statement.

R

```
uniquevals <- unique(z)
numunique <- length(uniquevals)
formula <- as.formula(y ~ x1 + ... + xk)
p <- length(coef(lm(formula)))
params <- matrix(rep(0, numunique*p), p, numunique)
for (i in 1:length(uniquevals)) {
    cat(i, "\n")
    params[,i] <- coef(lm(formula, subset=(z==uniquevals[i])))
}
```

Note: In the above code, separate regressions are fit for each value of the grouping variable be through use of a `for` loop. This requires the creation of a matrix of results `params` to be set up in advance, of the appropriate dimension (number of rows equal to the number of parameters (p=k+1) for the model, and number of columns equal to the number of levels for the grouping variable `z`). Within the loop, the `lm()` function is called and the coefficients from each fit are saved in the appropriate column of the `params` matrix.

3.1.7 One-way analysis of variance

HELP example: see 3.7.5

SAS

```
proc glm data=ds;
   class x;
   model y = x / solution;
run;
```

Note: The `solution` option to the `model` statement requests that the parameter estimates be displayed. Other procedures which fit ANOVA models include `proc anova` and `proc mixed`.

R

```
xf <- as.factor(x)
mod1 <- aov(y ~ xf, data=ds)
summary(mod1)
```

Note: The `summary()` command can be used to provide details of the model fit. More information can be found using `help(summary.aov)`. Note that `summary.lm(mod1)` will display the regression parameters underlying the ANOVA model.

3.1.8 Two-way (or more) analysis of variance

HELP example: see 3.7.5

Interactions can be specified using the syntax introduced in section 3.1.5 (see also interaction plots, section 5.1.9).

SAS

```
proc glm data=ds;
   class x1 x2;
   model y = x1 x2;
run;
```

Note: Other procedures which fit ANOVA models include `proc anova` and `proc mixed`.

R

```
aov(y ~ as.factor(x1) + as.factor(x2), data=ds)
```

3.2 Model comparison and selection

3.2.1 Compare two models

HELP example: see 3.7.5

Model comparison marks a key point of divergence for SAS and R. In general, most procedures in SAS fit a single model. Comparisons between models must be constructed by hand. An exception is "leave-one-out" models, in which a model identical to the one fit is considered, except that a single predictor is to be omitted. In this case, SAS offers "Type III" sums of squares tests, which can be printed by default or request in many modeling procedures. The R function `drop1()` computes a table of changes in fit. In addition, R offers functions which compare nested models using the `anova()` function. The Wald tests calculated by SAS and the likelihood ratio tests from `anova()` are identical in many settings, though they differ in general. In cases in which they differ, likelihood ratio tests are to be preferred.

R

```
mod1 <- lm(y ~ x1 + ... + xk, data=ds)
mod2 <- lm(y ~ x3 + ... + xk, data=ds)
anova(mod2, mod1)
```

or

```
drop1(mod2)
```

Note: The `anova()` command in R computes analysis of variance (or deviance) tables. When given one model as an argument, it displays the ANOVA table. When two (or more) nested models are given, it calculates the differences between them.

3.2.2 Log-likelihood

See also 3.2.3 (AIC) *HELP example:* see 3.7.5

SAS

```
proc mixed data=ds;
   model y = x1 ... xk;
run;
```

Note: Log-likelihood values are produced by various SAS procedures, but the means of requesting them can be idiosyncratic. The `mixed` procedure fits a superset of models available in `proc glm`, and can be used to generate this quantity.

R

```
mod1 <- lm(y ~ x1 + ... + xk, data=ds)
logLik(mod1)
```

Note: As of this writing, the `logLik()` function supports `glm`, `lm`, `nls`, `Arima`, `gls`, `lme`, and `nlme` objects.

3.2.3 Akaike Information Criterion (AIC)

See also 3.2.2 (log-likelihood) *HELP example:* see 3.7.5

SAS

```
proc reg data=ds stats=aic;
   model y = x1 ... xk;
run;
```

Note: AIC values are available in various SAS procedures, but the means of requesting them can be idiosyncratic.

R

```
mod1 <- lm(y ~ x1 + ... + xk, data=ds)
AIC(mod1)
```

Note: The `AIC()` function includes support for `glm`, `lm`, `nls`, `Arima`, `gls`, `lme`, and `nlme` objects.

3.2.4 Bayesian Information Criterion (BIC)

See also 3.2.3 (AIC)

SAS

```
proc mixed data=ds;
   model y = x1 ... xk;
run;
```

Note: BIC values are presented by default in `proc mixed`.

R

```
library(nlme)
mod1 <- lm(y ~ x1 + ... + xk, data=ds)
BIC(mod1)
```

3.3 Tests, contrasts, and linear functions of parameters

3.3.1 Joint null hypotheses: several parameters equal 0

SAS

```
proc reg data=ds;
   model ...;
   nametest: test varname1=0, varname2=0;
run;
```

Note: In the above, `nametest` is an arbitrary label which will appear in the output. Multiple `test` statements are permitted.

R

```
mod1 <- lm(y ~ x1 + ... + xk, data=ds)
mod2 <- lm(y ~ x3 + ... + xk, data=ds)
anova(mod2, mod1)
```

or

```
sumvals <- summary(mod1)
covb <- vcov(mod1)
coeff.mod1 <- coef(mod1)[2:3]
covmat <- matrix(c(covb[2,2], covb[2,3], covb[2,3], covb[3,3]), nrow=2)
fval <- t(coeff.mod1) %*% solve(covmat) %*% coeff.mod1
pval <- 1-pf(fval, 2, mod1$df)
```

Note: The R code for the second option, while somewhat complex, builds on the syntax introduced in 3.5.3, 3.5.8, and 3.5.9, and is intended to demonstrate ways to interact with linear model objects.

3.3.2 Joint null hypotheses: sum of parameters

SAS

```
proc reg data=ds;
   model ...;
   nametest: test varname1 + varname2=1;
run;
```

Note: The `test` statement is prefixed with an arbitrary `nametest` which will appear in the output. Multiple `test` statements are permitted.

R

```
mod1 <- lm(y ~ x1 + ... + xk, data=ds)
mod2 <- lm(y ~ I(x1+x2-1) + ... + xk, data=ds)
anova(mod2, mod1)
```

or

```
mod1 <- lm(y ~ x1 + ... + xk, data=ds)
covb <- vcov(mod1)
coeff.mod1 <- coef(mod1)
t <- (coeff.mod1[2,1]+coeff.mod1[3,1]-1)/
   sqrt(covb[2,2]+covb[3,3]+2*covb[2,3])
pvalue <- 2*(1-pt(abs(t), mod1$df))
```

Note: The `I()` function inhibits the interpretation of operators, to allow them to be used as arithmetic operators. The R code in the lower example utilizes the same approach introduced in 3.3.1.

3.3.3 Tests of equality of parameters

SAS *HELP example:* see 3.7.7

```
proc reg data=ds;
   model ...;
   nametest: test varname1=varname2;
run;
```

Note: The `test` statement is prefixed with an arbitrary `nametest` which will appear in the output. Multiple `test` statements are permitted.

R

```
mod1 <- lm(y ~ x1 + ... + xk, data=ds)
mod2 <- lm(y ~ I(x1+x2) + ... + xk, data=ds)
anova(mod2, mod1)
```

or

```
library(gmodels)
fit.contrast(mod1, "x1", values)
```

or

```
mod1 <- lm(y ~ x1 + ... + xk, data=ds)
covb <- vcov(mod1)
coeff.mod1 <- coef(mod1)
t <- (coeff.mod1[2]-coeff.mod1[3])/sqrt(covb[2,2]+covb[3,3]-2*covb[2,3])
pvalue <- 2*(1-pt(abs(t), mod1$df))
```

Note: The `I()` function inhibits the interpretation of operators, to allow them to be used as arithmetic operators. The `fit.contrast()` function calculates a contrast in terms of levels of the factor variable `x1` using a numeric matrix vector of contrast coefficients (where each row sums to zero) denoted by `values`. The more general R code below utilizes the same approach introduced in 3.3.1 for the specific test of $\beta_1 = \beta_2$ (different coding would be needed for other comparisons).

3.3.4 Multiple comparisons

SAS *HELP example:* see 3.7.6

```
proc glm data=ds;
   class x1;
   model y = x1;
   lsmeans x1 / pdiff adjust=tukey;
run;
```

Note: The `pdiff` option requests p-values for the hypotheses involving the pairwise comparison of means. The `adjust` option adjusts these p-values for multiple comparisons. Other options available through `adjust` include `bon` (for Bonferroni), and `dunnett`, among others. SAS `proc mixed` also has an `adjust` option for its `lsmeans` statement. A graphical presentation of significant differences among levels can be obtained with the `lines` option to the `lsmeans` statement, as shown in 3.7.6.

R

```
mod1 <- aov(y ~ x))
TukeyHSD(mod1, "x")
```

Note: The `TukeyHSD()` function takes an argument an `aov` object, and calculates the pairwise comparisons of all of the combinations of the factor levels of the variable `x` (see also `library(multcomp)`).

3.3.5 Linear combinations of parameters

HELP example: see 3.7.7

It is often useful to calculate predicted values for particular covariate values. Here, we calculate the predicted value $E[Y|X_1 = 1, X_2 = 3] = \hat{\beta}_0 + \hat{\beta}_1 + 3\hat{\beta}_2$.

SAS

```
proc glm data=ds;
   model y = x1 ... xk;
   estimate 'label' intercept 1 x1 1 x2 3;
run;
```

Note: The `estimate` statement is used to calculate linear combination of parameters (and associated standard errors). The optional quoted text is a label which will be printed with the estimated function.

R

```
newdf <- data.frame(x1=c(1), x2=c(3))
estimates <- predict(mod1, newdf, se.fit=TRUE, interval="confidence")
```

Note: The `predict()` command in R can generate estimates at any combination of parameter values, as specified as a dataframe that is passed as an argument. More information on this function can be found using `help(predict.lm)`.

3.4 Model diagnostics

3.4.1 Predicted values

HELP example: see 3.7.2

SAS

```
proc reg data=ds;
   model ...;
   output out=newds predicted=predicted_varname;
run;
```

or

```
proc glm data=ds;
   model ...;
   output out=newds predicted=predicted_varname;
run;
```

Note: The `output` statement creates a new dataset and specifies variables to be included, of which the predicted values are an example. Others can be found using the on-line help: Contents; SAS Products; SAS Procedures; REG; Output Statement.

R

```
mod1 <- lm(...)
predicted.varname <- predict(mod1)
```

Note: The command `predict()` operates on any `lm()` object, and by default generates a vector of predicted values. Similar commands retrieve other regression output.

3.4.2 Residuals

SAS *HELP example:* see 3.7.2

```
proc glm data=ds;
   model ...;
   output out=newds residual=residual_varname;
run;
```

or

```
proc reg data=ds;
   model ...;
   output out=newds residual=residual_varname;
run;
```

Note: The `output` statement creates a new dataset and specifies variables to be included, of which the residuals are an example. Others can be found using the on-line help: Contents; SAS Products; SAS Procedures; Proc REG; Output Statement.

R

```
mod1 <- lm(...)
residual.varname <- residuals(mod1)
```

Note: The command `residuals()` operates on any `lm()` object, and generates a vector of residuals. Other functions for analysis of variance objects, GLM or linear mixed effects exist (see for example `help(residuals.glm)`).

3.4.3 Studentized residuals

HELP example: see 3.7.2

Standardized residuals are calculated by dividing the ordinary residual (observed minus expected, $y_i - \hat{y}_i$) by an estimate of its standard deviation. Studentized residuals are calculated in a similar manner, where the predicted value and the variance of the residual are estimated from the model fit while excluding that observation. In SAS `proc glm` the standardized residual is requested by the `student` option, while the `rstudent` option generates the studentized residual.

SAS

```
proc glm data=ds;
   model ...;
   output out=newds student=standardized_resid_varname;
run;
```

or

```
proc reg data=ds;
   model ...;
   output out=newds rstudent=studentized_resid_varname;
run;
```

Note: The output statement creates a new dataset and specifies variables to be included, of which the studentized residuals are an example. Both proc reg and proc glm include both types of residuals. Others can be found using the on-line help: Contents; SAS Products; SAS Procedures; Proc REG; Output Statement.

R

```
mod1 <- lm(...)
standardized.resid.varname <- stdres(mod1)
studentized.resid.varname <- studres(mod1)
```

Note: The stdres() and studres() functions operate on any lm() object, and generate a vector of studentized residuals (the former command includes the observation in the calculation, while the latter does not). Similar commands retrieve other regression output (see help(influence.measures)).

3.4.4 Leverage

HELP example: see 3.7.2

Leverage is defined as the diagonal element of the $(X(X^TX)^{-1}X^T)$ or "hat" matrix.

SAS

```
proc glm data=ds;
   model ...;
   output out=newds h=leverage_varname;
run;
```

or

```
proc reg data=ds;
   model ...;
   output out=newds h=leverage_varname;
run;
```

Note: The output statement creates a new dataset and specifies variables to be included, of which the leverage values are one example. Others can be found using the on-line help: Contents; SAS Products; SAS Procedures; Proc REG; Output Statement.

R

```
mod1 <- lm(...)
leverage.varname <- hatvalues(mod1)
```

Note: The command hatvalues() operates on any lm() object, and generates a vector of leverage values. Similar commands can be utilized to retrieve other regression output (see help(influence.measures)).

3.4.5 Cook's D

HELP example: see 3.7.2

Cook's distance (D) is a function of the leverage (see 3.4.4) and the residual. It is used as a measure of the influence of a data point in a regression model.

SAS

```
proc glm data=ds;
   model ...;
   output out=newds cookd=cookd_varname;
run;
```

or

```
proc reg data=ds;
   model ...;
   output out=newds cookd=cookd_varname;
run;
```

Note: The `output` statement creates a new dataset and specifies variables to be included, of which the Cook's distance values are an example. Others can be found using the on-line help: Contents; SAS Products; SAS Procedures; Proc REG; Output Statement.

R

```
mod1 <- lm(...)
cookd.varname <- cooks.distance(mod1)
```

Note: The command `cooks.distance()` operates on any `lm()` object, and generates a vector of Cook's distance values. Similar commands retrieve other regression output.

3.4.6 DFFITS

HELP example: see 3.7.2

DFFITS are a standardized function of the difference between the predicted value for the observation when it is included in the dataset and when (only) it is excluded from the dataset. They are used as an indicator of the observation's influence.

SAS

```
proc reg data=ds;
   model ...;
   output out=newds dffits=dffits_varname;
run;
```

or

```
proc glm data=ds;
   model ...;
   output out=newds dffits=dffits_varname;
run;
```

Note: The `output` statement creates a new dataset and specifies variables to be included, of which the dffits values are an example. Others can be found using the on-line help: Contents; SAS Products; SAS Procedures; Proc REG; Output Statement.

R

```
mod1 <- lm(...)
dffits.varname <- dffits(mod1)
```

Note: The command `dffits()` operates on any `lm()` object, and generates a vector of dffits values. Similar commands retrieve other regression output.

3.4.7 Diagnostic plots

HELP example: see 3.7.3

SAS

```
proc reg data=ds;
   model ...
   output out=newds predicted=pred_varname residual=resid_varname
          h=leverage_varname cookd=cookd_varname;
run;

proc gplot data=ds;
   plot resid_varname * pred_varname;
   plot resid_varname * leverage_varname;
run;
quit;
```

Note: To mimic R more closely, use a data step to generate the square root of residuals. QQ plots of residuals can be generated via proc univariate. It is not straightforward to plot lines of constant Cook's D on the residuals vs. leverage plot. The `ods graphics on` statement (A.7.3), issued prior to running the `reg` procedure will produce many diagnostic plots, as will running `ods graphics on` and then `proc glm` with the `plots=diagnostics` option.

R

```
mod1 <- lm(...)
par(mfrow=c(2, 2)) # display 2 x 2 matrix of graphs
plot(mod1)
```

Note: The `plot.lm()` function (which is invoked when `plot()` is given a linear regression model as an argument) can generate six plots: 1) a plot of residuals against fitted values, 2) a Scale-Location plot of $\sqrt{(}Y_i - \hat{Y}_i)$ against fitted values, 3) a Normal Q-Q plot of the residuals, 4) a plot of Cook's distances (3.4.5) versus row labels, 5) a plot of residuals against leverages (3.4.4), and 6) a plot of Cook's distances against leverage/(1-leverage). The default is to plot the first three and the fifth. The `which` option can be used to specify a different set (see `help(plot.lm)`).

3.5 Model parameters and results

3.5.1 Prediction limits

These are the lower (and upper) prediction limits for 'new' observations with the covariate values of subjects observed in the dataset, as opposed to confidence limits for the population mean (see 3.5.4).

SAS

```
proc glm data=ds;
   model ...;
   output out=newds lcl=lcl_varname;
run;
```

or

```
proc reg data=ds;
   model ...;
   output out=newds lcl=lcl_varname;
run;
```

Note: The `output` statement creates a new dataset and specifies variables to be included, of which the lower prediction limit values are an example. The upper limits can be requested with the `ucl` option to the `output` statement. Other possibilities can be found using the on-line help: Contents; SAS Products; SAS Procedures; Proc REG; Output Statement.

R

```
mod1 <- lm(y ~ ..., data=ds)
pred.w.lowlim <- predict(mod1, interval="prediction")[,2]
```

Note: This code saves the second column of the results from the `predict()` function into a vector. To generate the upper confidence limits, the user would access the third column of the `predict()` object in R. The command `predict()` operates on any `lm()` object, and with these options generates prediction limit values. By default, the function uses the estimation dataset, but a separate dataset of values to be used to predict can be specified.

3.5.2 Parameter estimates

HELP example: see 3.7.2

SAS

```
ods output parameterestimates=newds;
proc glm data=ds;
   model ... / solution;
run;
```

or

```
proc reg data=ds outest=newds;
   model ...;
run;
```

Note: The `ods output` statement (section A.7.1) can be used to save any piece of SAS output as a SAS dataset. The `outest` option is specific to `proc reg`, though many other procedures accept similar syntax.

R

```
mod1 <- lm(...)
coeff.mod1 <- coef(mod1)
```

Note: The first element of the vector `coeff.mod1` is the intercept (assuming that a model with an intercept was fit).

3.5.3 Standard errors of parameter estimates

See also 3.5.9 (covariance matrix)

SAS

```
proc reg data=ds outest=newds;
   model .../ outseb ...;
run;
```

or

```
ods output parameterestimates=newds;
proc glm data=ds;
   model .../ solution;
run;
```

Note: The `ods output` statement (section A.7.1) can be used to save any piece of SAS output as a SAS dataset.

R

```
mod1 <- lm(...)
se.mod1 <- coef(summary(mod1))[,2]
```

Note: The standard errors are the second column of the results from `coef()`.

3.5.4 Confidence limits for the mean

These are the lower (and upper) confidence limits for the mean of observations with the given covariate values, as opposed to the prediction limits for individual observations with those values (see 3.5.1).

SAS

```
proc glm data=ds;
   model ...;
   output out=newds lclm=lcl_mean_varname;
run;
```

or

```
proc reg data=ds;
   model ...;
   output out=newds lclm=lcl_mean_varname;
run;
```

Note: The `output` statement creates a new dataset and specifies output variables to be included, of which the lower confidence limit values are one example. The upper confidence limits can be generated using the `uclm` option to the `output` statement. Other possibilities can be found using the on-line help: Contents; SAS Products; SAS Procedures; Proc REG; Output Statement.

R

```
mod1 <- lm(...)
pred <- predict(mod1, interval="confidence")
lcl.varname <- pred[,2]
```

Note: The lower confidence limits are the second column of the results from `coef()`. To generate the upper confidence limits, the user would replace `lclm` with `uclm` for SAS and access the third column of the `predict()` object in R. The command `predict()` operates on any `lm()` object, and with these options generates confidence limit values. By default, the function uses the estimation dataset, but a separate dataset of values to be used to predict can be specified.

3.5.5 Plot confidence intervals for the mean

SAS

```
symbol1 i=rlclm95 value=none;
proc gplot data=ds;
   plot y * x;
run;
```

Note: The symbol statement i option (synonym for interpolation) contains many useful options for adding features to scatterplots. The rlclm95 selection requests a regression line plot, with 95% confidence limits for the mean. The value=none requests that the observations themselves not be plotted (see also scatterplots, 5.1.1).

R

```
pred.w.clim <- predict(lm(y ~ x), interval="confidence")
matplot(new$x, pred.w.clim, lty=c(1, 2, 2), type="l", ylab="predicted y")
```

Note: This entry produces fit and confidence limits at the original observations in the original order. If the observations aren't sorted relative to the explanatory variable x, the resulting plot will be a jumble. The matplot() function is used to generate lines, with a solid line (lty=1) for predicted values and dashed line (lty=2) for the confidence bounds.

3.5.6 Plot prediction limits from a simple linear regression

SAS

```
symbol1 i=rlcli95 l=2 value=none;
proc gplot data=ds;
   plot y * x;
run;
```

Note: The symbol statement i (synonym for interpolation) option contains many useful options for adding features to scatterplots (see also 5.1.1). The rlcli95 selection requests a regression line plot, with 95% confidence limits for the values. The value=none requests that the observations not be plotted.

R

```
pred.w.plim <- predict(lm(y ~ x), interval="prediction")
matplot(new$x, pred.w.plim, lty=c(1, 2, 2), type="l", ylab="predicted y")
```

Note: This entry produces fit and confidence limits at the original observations in the original order. If the observations aren't sorted relative to the explanatory variable x, the resulting plot will be a jumble. The matplot() function is used to generate lines, with a solid line (lty=1) for predicted values and dashed line (lty=2) for the confidence bounds.

3.5.7 Plot predicted lines for each value of a variable

Here we describe how to generate plots for a variable X_1 versus Y separately for each value of the variable X_2 (see also conditioning plot, 5.1.11).

SAS

```
symbol1 i=rl value=none;
symbol2 i=rl value=none;
proc gplot data=ds;
   plot y*x1 = x2;
run;
```

Note: The symbol statement i (synonym for interpolation) option contains many useful options for adding features to scatterplots. The rl selection requests a regression line plot.

The `value=none` requests that the observations not be plotted. The `= x2` syntax requests a different `symbol` statement be applied for each level of `x2` (see also scatterplots, 5.1.1).

R

```
plot(x, y, pch=" ") # create an empty plot of the correct size
abline(lm(y ~ x1, subset=x2==0), lty=1, lwd=2)
abline(lm(y ~ x1, subset=x2==1), lty=2, lwd=2)
...
abline(lm(y ~ x1, subset=x2==k), lty=k+1, lwd=2)
```

Note: The `abline()` function is used to generate lines for each of the subsets, with a solid line (`lty=1`) for the first group and dashed line (`lty=2`) for the second (this assumes that X_2 takes on values 0–k, see 3.1.6).

3.5.8 Design and information matrix

See also 1.9 (matrices)

SAS

```
proc reg data=ds;
   model .../ xpx ...;
run;
```

or

```
proc glm data=ds;
   model .../ xpx ...;
run;
```

Note: A dataset containing the information ($X'X$) matrix can be created using `ODS` by specifying either `proc` statement or by adding the option `outsscp=newds` to the `proc reg` statement.

R

```
mod1 <- lm(y ~ x1 + ... + xk, data=ds)
XpX <- t(model.matrix(mod1)) %*% model.matrix(mod1)
```

or

```
X <- cbind(rep(1, length(x1)), x1, x2, ..., xk)
XpX <- t(X) %*% X
rm(X)
```

Note: The `model.matrix()` function creates the design matrix from a linear model object. Alternatively, this quantity can be built up using the `cbind()` function to glue together the design matrix `X`. Finally, matrix multiplication and the transpose function are used to create the information ($X'X$) matrix.

3.5.9 Covariance matrix

See also 1.9 (matrices) and 3.5.3 (standard errors)

HELP example: see 3.7.2

SAS

```
proc reg data=ds outest=newds covout;
run;
```

or

```
ods output covb=newds;
proc reg data=ds;
   model ... / covb ...;
```

R

```
mod1 <- lm(...)
varcov <- vcov(mod1)
```

or

```
sumvals <- summary(mod1)
covb <- sumvals$cov.unscaled*sumvals$sigma^2
```

Note: Running `help(summary.lm)` provides details on return values.

3.6 Further resources

Accessible guides to linear regression in R and SAS can be found in [19] and [49], respectively. Cook [14] reviews regression diagnostics. The CRAN Task View on Statistics for the Social Sciences provides an excellent overview of methods described here and in Chapter 4.

3.7 HELP examples

To help illustrate the tools presented in this chapter, we apply many of the entries to the HELP data. SAS and R code can be downloaded from `http://www.math.smith.edu/sasr/examples`.

We begin by reading in the dataset and keeping only the female subjects. In R, we create a version of the `substance` variable as a factor (see 3.1.3).

```
proc import datafile='c:/book/help.dta'
   out=help_a dbms=dta;
run;

data help;
set help_a;
   if female;
run;
```

```
> options(digits=3)
> # read in Stata format
> library(foreign)
> ds <- read.dta("help.dta", convert.underscore=FALSE)
> newds <- ds[ds$female==1,]
> attach(newds)
> sub <- factor(substance, levels=c("heroin", "alcohol", "cocaine"))
```

3.7.1 Scatterplot with smooth fit

As a first step to help guide fitting a linear regression, we create a scatterplot (5.1.1) displaying the relationship between age and the number of alcoholic drinks consumed in the period before entering detox (variable name: `i1`), as well as primary substance of abuse (alcohol, cocaine or heroin).

Figure 3.7.1 displays a scatterplot of observed values for `i1` (along with separate smooth fits by primary substance). To improve legibility, the plotting region is restricted to those with number of drinks between 0 and 40 (see plotting limits, 5.3.7).

```
axis1 order = (0 to 40 by 10) minor=none;
axis2 minor=none;
legend1 label=none value=(h=1.5) shape=symbol(10,1.2)
   down=3 position=(top right inside) frame mode=protect;
symbol1 v=circle i=sm70s c=black l=1 h=1.1 w=5;
symbol2 v=diamond i=sm70s c=black l=33 h=1.1 w=5;
symbol3 v=square i=sm70s c=black l=8 h=1.1 w=5;
proc gplot data=help;
   plot i1*age = substance / vaxis=axis1 haxis=axis2 legend=legend1;
run; quit;
```

```
> plot(age, i1, ylim=c(0,40), type="n", cex.lab=1.4, cex.axis=1.4)
> points(age[substance=="alcohol"], i1[substance=="alcohol"], pch="a")
> lines(lowess(age[substance=="alcohol"],
+    i1[substance=="alcohol"]), lty=1, lwd=2)
> points(age[substance=="cocaine"], i1[substance=="cocaine"], pch="c")
> lines(lowess(age[substance=="cocaine"],
+    i1[substance=="cocaine"]), lty=2, lwd=2)
> points(age[substance=="heroin"], i1[substance=="heroin"], pch="h")
> lines(lowess(age[substance=="heroin"],
+    i1[substance=="heroin"]), lty=3, lwd=2)
> legend(44, 38, legend=c("alcohol", "cocaine", "heroin"), lty=1:3,
+    cex=1.4, lwd=2, pch=c("a", "c", "h"))
```

The `pch` option to the `legend()` command can be used to insert plot symbols in R legends (Figure 3.1 displays the different line-styles).

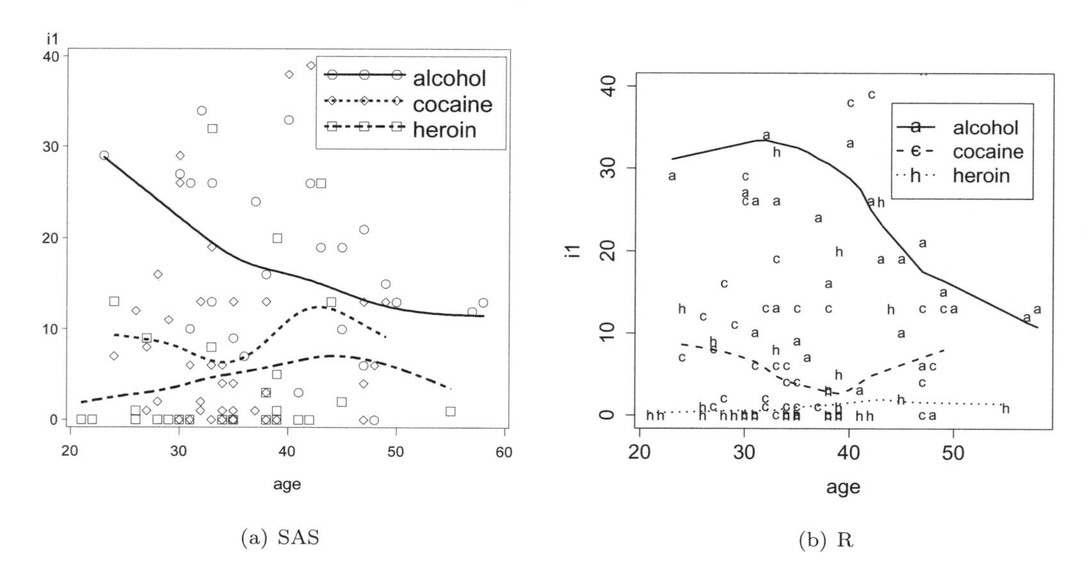

(a) SAS (b) R

Figure 3.1: Scatterplot of observed values for AGE and I1 (plus smoothers by substance)

Not surprisingly, Figure 3.1 suggests that there is a dramatic effect of primary substance, with alcohol users drinking more than others. There is some indication of an interaction with age. It is important to note that SAS uses only the points displayed (i.e. within the specified axes) when smoothing, while R uses all points, regardless of whether they appear in the plot.

3.7.2 Linear regression with interaction

Next we fit a linear regression model (3.1.1) for the number of drinks as a function of age, substance, and their interaction (3.1.5). To assess the need for the interaction, we use the F test from the Type III sums of squares in SAS. In R, we additionally fit the model with no interaction and use the `anova()` function to compare the models (the `drop1()` function could also be used). To save space, some results of `proc glm` have been suppressed using the `ods select` statement (see A.7).

```
options ls=74;  /* reduces width of output to make it fit in gray area */
ods select overallanova modelanova parameterestimates;
proc glm data=help;
class substance;
   model i1 = age substance age * substance / solution;
   output out=helpout cookd=cookd_ch4 dffits=dffits_ch4
     student=sresids_ch4 residual=resid_ch4
     predicted=pred_ch4 h=lev_ch4;
run; quit;
ods select all;
```

```
The GLM Procedure

Dependent Variable: I1   i1

                                   Sum of
Source                      DF    Squares    Mean Square  F Value  Pr > F

Model                        5  12275.17570   2455.03514     9.99  <.0001

Error                      101  24815.36635    245.69670

Corrected Total            106  37090.54206
The GLM Procedure

Dependent Variable: I1   i1

Source                      DF    Type I SS  Mean Square  F Value  Pr > F

AGE                          1    384.75504    384.75504     1.57  0.2137
SUBSTANCE                    2  10509.56444   5254.78222    21.39  <.0001
AGE*SUBSTANCE                2   1380.85622    690.42811     2.81  0.0649

Source                      DF  Type III SS  Mean Square  F Value  Pr > F
```

```
AGE                          1      27.157727       27.157727       0.11   0.7402
SUBSTANCE                    2    3318.992822     1659.496411       6.75   0.0018
AGE*SUBSTANCE                2    1380.856222      690.428111       2.81   0.0649

The GLM Procedure

Dependent Variable: I1    i1

                                                 Standard
Parameter                   Estimate                Error   t Value   Pr > |t|

Intercept               -7.77045212 B      12.87885672     -0.60     0.5476
AGE                      0.39337843 B       0.36221749      1.09     0.2801
SUBSTANCE     alcohol   64.88044165 B      18.48733701      3.51     0.0007
SUBSTANCE     cocaine   13.02733169 B      19.13852222      0.68     0.4976
SUBSTANCE     heroin     0.00000000 B        .               .        .
AGE*SUBSTANCE alcohol   -1.11320795 B       0.49135408     -2.27     0.0256
AGE*SUBSTANCE cocaine   -0.27758561 B       0.53967749     -0.51     0.6081
AGE*SUBSTANCE heroin     0.00000000 B        .               .        .
```

```
> options(show.signif.stars=FALSE)
> lm1 <- lm(i1 ~ sub * age)
> lm2 <- lm(i1 ~ sub + age)
> anova(lm2, lm1)

Analysis of Variance Table

Model 1: i1 ~ sub + age
Model 2: i1 ~ sub * age
  Res.Df   RSS  Df Sum of Sq     F Pr(>F)
1    103 26196
2    101 24815   2      1381 2.81  0.065
```

There is some indication of a borderline significant interaction between age and substance group (p=0.065).

In SAS, the `ods output` statement can be used to save any printed result as a SAS dataset. In the following code, all printed output from `proc glm` is suppressed, but the parameter estimates are saved as a SAS dataset, then printed using `proc print`. In addition, various diagnostics are saved via the the `output` statement.

```
ods select none;
ods output parameterestimates=helpmodelanova;
proc glm data=help;
class substance;
model i1 = age|substance / solution;
output out=helpout cookd=cookd_ch4 dffits=dffits_ch4
    student=sresids_ch4 residual=resid_ch4
    predicted=pred_ch4 h=lev_ch4;
run; quit;
ods select all;
```

```
proc print data=helpmodelanova;
   var parameter estimate stderr tvalue probt;
   format _numeric_ 6.3;
run;

Obs    Parameter                    Estimate    StdErr    tValue     Probt

 1     Intercept                     -7.770     12.879    -0.603     0.548
 2     AGE                            0.393      0.362     1.086     0.280
 3     SUBSTANCE      alcohol        64.880     18.487     3.509     0.001
 4     SUBSTANCE      cocaine        13.027     19.139     0.681     0.498
 5     SUBSTANCE      heroin          0.000       .          .         .
 6     AGE*SUBSTANCE alcohol         -1.113      0.491    -2.266     0.026
 7     AGE*SUBSTANCE cocaine         -0.278      0.540    -0.514     0.608
 8     AGE*SUBSTANCE heroin           0.000       .          .         .
```

In R, we can get similar information with the `summary()` function.

```
> summary(lm1)

Call:
lm(formula = i1 ~ sub * age)

Residuals:
   Min     1Q Median     3Q    Max
-31.92  -8.25  -4.18   3.58  49.88

Coefficients:
               Estimate Std. Error t value Pr(>|t|)
(Intercept)      -7.770     12.879   -0.60  0.54763
subalcohol       64.880     18.487    3.51  0.00067
subcocaine       13.027     19.139    0.68  0.49763
age               0.393      0.362    1.09  0.28005
subalcohol:age   -1.113      0.491   -2.27  0.02561
subcocaine:age   -0.278      0.540   -0.51  0.60813

Residual standard error: 15.7 on 101 degrees of freedom
Multiple R-squared: 0.331,      Adjusted R-squared: 0.298
F-statistic: 9.99 on 5 and 101 DF,  p-value: 8.67e-08
```

There are many quantities of interest stored in the linear model object `lm1`, and these can be viewed or extracted for further use.

```
> names(summary(lm1))

 [1] "call"          "terms"         "residuals"     "coefficients"
 [5] "aliased"       "sigma"         "df"            "r.squared"
 [9] "adj.r.squared" "fstatistic"    "cov.unscaled"

> summary(lm1)$sigma

[1] 15.7
```

```
> names(lm1)

 [1] "coefficients" "residuals"     "effects"       "rank"
 [5] "fitted.values" "assign"       "qr"            "df.residual"
 [9] "contrasts"    "xlevels"       "call"          "terms"
[13] "model"
```

```
> lm1$coefficients

   (Intercept)      subalcohol      subcocaine             age subalcohol:age
        -7.770          64.880          13.027           0.393         -1.113
subcocaine:age
        -0.278

> coef(lm1)

   (Intercept)      subalcohol      subcocaine             age subalcohol:age
        -7.770          64.880          13.027           0.393         -1.113
subcocaine:age
        -0.278
```

```
> vcov(lm1)

               (Intercept) subalcohol subcocaine    age subalcohol:age
(Intercept)         165.86    -165.86    -165.86 -4.548          4.548
subalcohol         -165.86     341.78     165.86  4.548         -8.866
subcocaine         -165.86     165.86     366.28  4.548         -4.548
age                  -4.55       4.55       4.55  0.131         -0.131
subalcohol:age        4.55      -8.87      -4.55 -0.131          0.241
subcocaine:age        4.55      -4.55     -10.13 -0.131          0.131
               subcocaine:age
(Intercept)             4.548
subalcohol             -4.548
subcocaine            -10.127
age                    -0.131
subalcohol:age          0.131
subcocaine:age          0.291
```

3.7.3 Regression diagnostics

Assessing the model is an important part of any analysis. We begin by examining the residuals (3.4.2). First, we calculate the quantiles of their distribution, then display the smallest residual.

```
options ls=74;
proc means data=helpout min q1 median q3 max maxdec=2;
   var resid_ch4;
run;

The MEANS Procedure
```

```
                     Analysis Variable : resid_ch4

                        Lower                            Upper
     Minimum         Quartile         Median          Quartile          Maximum
-------------------------------------------------------------------------------
      -31.92            -8.31          -4.18              3.69            49.88
-------------------------------------------------------------------------------
```

```
> pred <- fitted(lm1)
> resid <- residuals(lm1)
> quantile(resid)

    0%     25%     50%     75%    100%
-31.92  -8.25  -4.18   3.58  49.88
```

We could examine the output, then condition to find the value of the residual that is less than -31. Instead the dataset can be sorted so the smallest observation is first and then print one observation.

```
proc sort data=helpout;
   by resid_ch4;
run;

proc print data=helpout (obs=1);
   var id age i1 substance pred_ch4 resid_ch4;
run;

                                                           resid_
Obs      ID      AGE      I1      SUBSTANCE      pred_ch4      ch4

  1     325      35        0       alcohol       31.9160    -31.9160
```

One way to print the largest value is to sort the dataset in the reverse order, then print just the first observation.

```
proc sort data=helpout;
   by descending resid_ch4;
run;

proc print data=helpout (obs=1);
   var id age i1 substance pred_ch4 resid_ch4;
run;

                                                         resid_
Obs      ID      AGE      I1      SUBSTANCE      pred_ch4      ch4

  1       9       50      71       alcohol       21.1185     49.8815
```

```
> tmpds <- data.frame(id, age, i1, sub, pred, resid, rstandard(lm1))
> tmpds[resid==max(resid),]

  id age i1     sub pred resid rstandard.lm1.
4  9  50 71 alcohol 21.1  49.9           3.32

> tmpds[resid==min(resid),]

    id age i1     sub pred resid rstandard.lm1.
72 325  35  0 alcohol 31.9 -31.9          -2.07
```

The R output includes the row number of the minimum and maximum residual.

Graphical tools are the best way to examine residuals. Figure 3.2 displays the default diagnostic plots (3.4) from the model (for R) and the Q-Q plot generated from the saved diagnostics (for SAS).

Sometimes in SAS it is necessary to clear out old graphics settings. This is easiest to do with the `goptions reset=all` statement (5.3.5).

```
goptions reset=all;
```

```
ods select univar;
proc univariate data=helpout;
   qqplot resid_ch4 / normal(mu=est sigma=est color=black);
run;
ods select all;
```

```
> oldpar <- par(mfrow=c(2, 2), mar=c(4, 4, 2, 2)+.1)
> plot(lm1)
> par(oldpar)
```

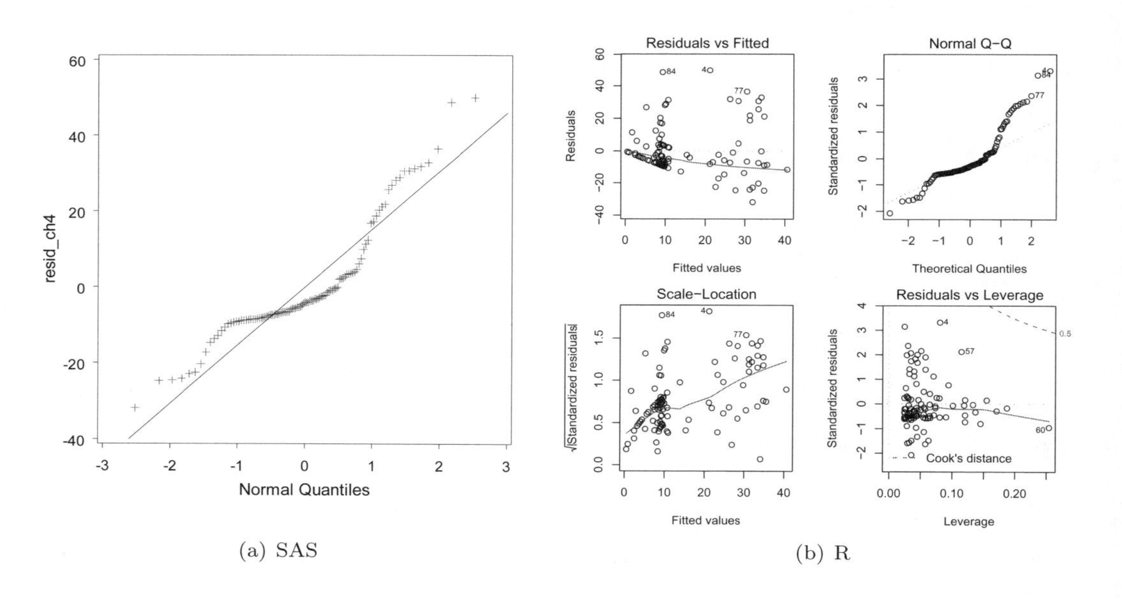

(a) SAS (b) R

Figure 3.2: Q-Q plot from SAS, default diagnostics from R

In SAS, we could use the `ods graphics` option to get assorted diagnostic plots, but here we demonstrate a manual approach using the previously saved diagnostics. Figure 3.3 displays the empirical density of the standardized residuals, along with an overlaid normal density. The assumption that the residuals are approximately Gaussian does not appear to be tenable.

```
axis1 label=("Standardized residuals");
ods select "Histogram 1";
proc univariate data=helpout;
   var sresids_ch4;
   histogram sresids_ch4 / normal(mu=est sigma=est color=black)
      kernel(color=black) haxis=axis1;
run;
ods select all;
```

```
> library(MASS)
> std.res <- rstandard(lm1)
> hist(std.res, breaks=seq(-2.5, 3.5, by=.5), main="",
+    xlab="standardized residuals", col="gray80", freq=FALSE)
> lines(density(std.res), lwd=2)
> xvals <- seq(from=min(std.res), to=max(std.res), length=100)
> lines(xvals, dnorm(xvals, mean(std.res), sd(std.res)), lty=2)
```

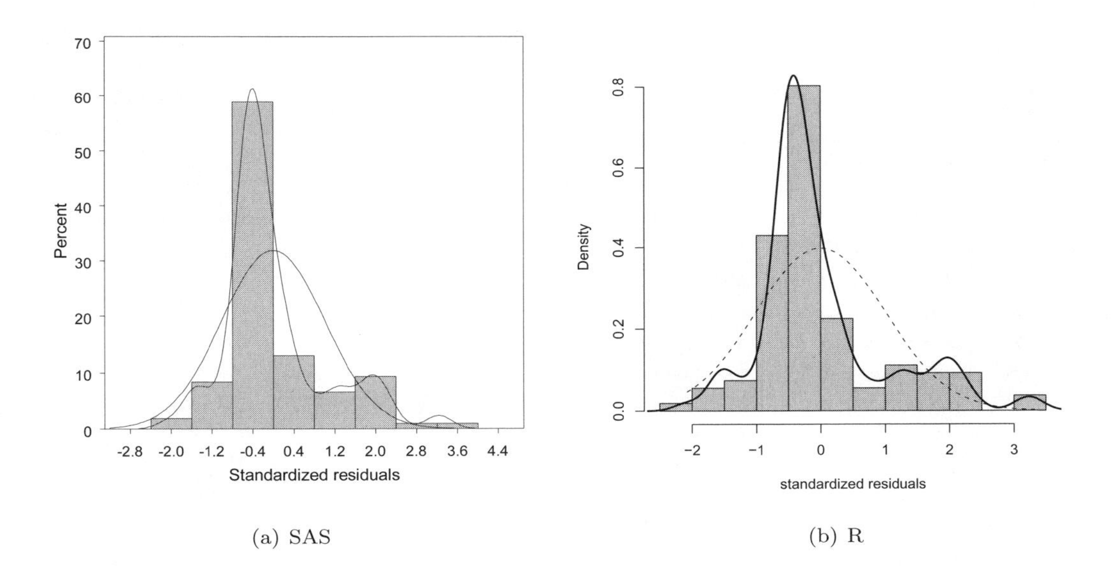

(a) SAS

(b) R

Figure 3.3: Empirical density of residuals, with superimposed normal density

The residual plots indicate some potentially important departures from model assumptions, and further exploration should be undertaken.

3.7.4 Fitting regression model separately for each value of another variable

One common task is to perform identical analyses in several groups. Here, as an example, we consider separate linear regressions for each substance abuse group. In SAS, we show only the parameter estimates, using `ODS`.

```
ods select none;
proc sort data=help;
   by substance;
run;

ods output parameterestimates=helpsubstparams;
proc glm data=help;
   by substance;
   model i1 = age / solution;
run;
ods select all;
```

```
options ls=74;
proc print data=helpsubstparams;
run;

Obs SUBSTANCE Dependent Parameter     Estimate       StdErr  tValue  Probt

 1   alcohol      I1     Intercept  57.10998953  18.00474934    3.17 0.0032
 2   alcohol      I1     AGE        -0.71982952   0.45069028   -1.60 0.1195
 3   cocaine      I1     Intercept   5.25687957  11.52989056    0.46 0.6510
 4   cocaine      I1     AGE         0.11579282   0.32582541    0.36 0.7242
 5   heroin       I1     Intercept  -7.77045212   8.59729637   -0.90 0.3738
 6   heroin       I1     AGE         0.39337843   0.24179872    1.63 0.1150
```

For R, a matrix of the correct size is created, then a `for` loop is run for each unique value of the grouping variable.

```
> uniquevals <- unique(substance)
> numunique <- length(uniquevals)
> formula <- as.formula(i1 ~ age)
> p <- length(coef(lm(formula)))
> res <- matrix(rep(0, numunique*p), p, numunique)
> for (i in 1:length(uniquevals)) {
+    res[,i] <- coef(lm(formula, subset=substance==uniquevals[i]))
+ }
> rownames(res) <- c("intercept","slope")
> colnames(res) <- uniquevals
> res

          heroin cocaine alcohol
intercept -7.770   5.257   57.11
slope      0.393   0.116   -0.72

> detach(newds)
```

3.7.5 Two way ANOVA

Is there a statistically significant association between gender and substance abuse group with depressive symptoms? In SAS, we can make an interaction plot (5.1.9) by hand, as below, or `proc glm` will make one automatically if the `ods graphics on` statement is issued.

```
libname k 'c:/book';

proc sort data=k.help;
   by substance female;
run;

ods select none;
proc means data=k.help;
   by substance female;
   var cesd;
   output out=helpmean mean=;
run;
ods select all;
```

```
axis1 minor=none;
symbol1 i=j v=none l=1 c=black w=5;
symbol2 i=j v=none l=2 c=black w=5;
proc gplot data=helpmean;
   plot cesd*substance = female / haxis=axis1 vaxis=axis1;
run; quit;
```

R has a function `interaction.plot()` to carry out this task. Figure 3.4 displays an interaction plot for CESD as a function of substance group and gender.

```
> attach(ds)
> sub <- as.factor(substance)
> genf <- as.factor(ifelse(female, "F", "M"))
> interaction.plot(sub, genf, cesd, xlab="substance", las=1, lwd=2)
```

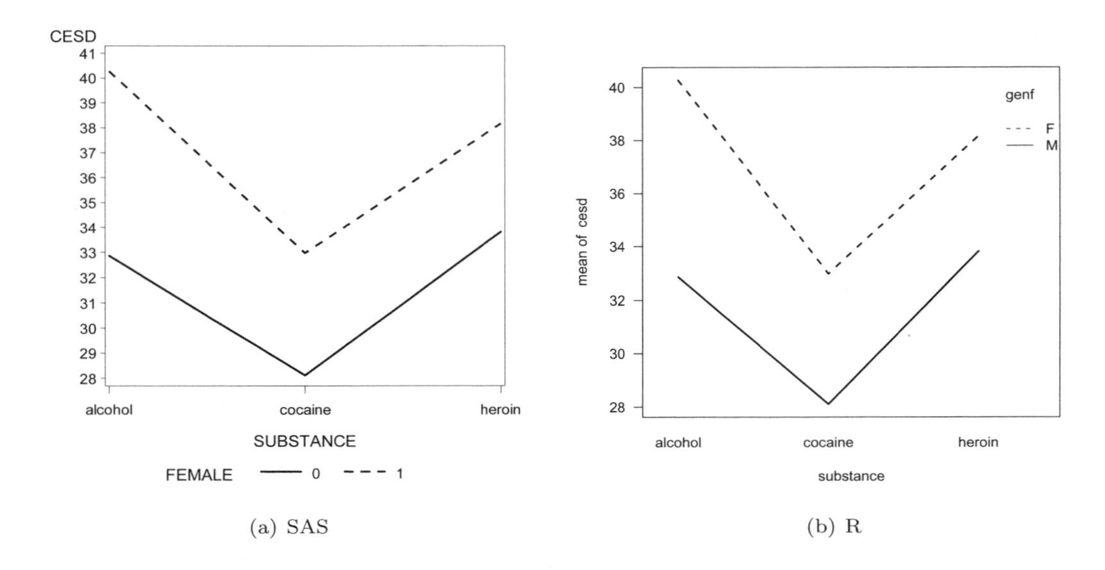

(a) SAS (b) R

Figure 3.4: Interaction plot of CESD as a function of substance group and gender

There are indications of large effects of gender and substance group, but little suggestion of interaction between the two. The same conclusion is reached in Figure 3.5, which displays boxplots by substance group and gender.

```
data h2; set k.help;
   if female eq 1 then sex='F';
   else sex='M';
run;

proc sort data=h2; by sex; run;
symbol1 v='x' c=black;
proc boxplot data=h2;
   plot cesd * substance(sex) / notches boxwidthscale=1;
run;
```

```
> subs <- character(length(substance))
> subs[substance=="alcohol"] <- "Alc"
> subs[substance=="cocaine"] <- "Coc"
> subs[substance=="heroin"] <- "Her"
> gen <- character(length(female))
> boxout <- boxplot(cesd ~ subs + genf, notch=TRUE, varwidth=TRUE,
+     col="gray80")
> boxmeans <- tapply(cesd, list(subs, genf), mean)
> points(seq(boxout$n), boxmeans, pch=4, cex=2)
```

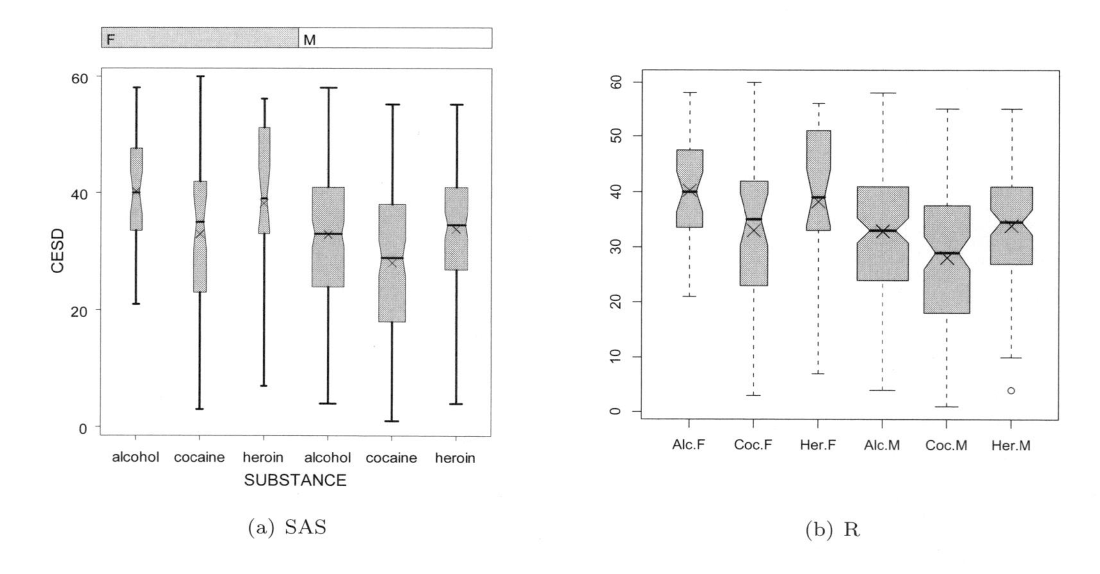

(a) SAS (b) R

Figure 3.5: Boxplot of CESD as a function of substance group and gender

The width of each box is proportional to the size of the sample, with the notches denoting confidence intervals for the medians, and X's marking the observed means.

Next, we proceed to formally test whether there is a significant interaction through a two-way analysis of variance (3.1.8). In SAS, the Type III sums of squares table can be used to assess the interaction; we restrict output to this table to save space. In R we fit models with and without an interaction, and then compare the results. We also construct the likelihood ratio test manually.

```
options ls=74;
ods select modelanova;
proc glm data=k.help;
   class female substance;
   model cesd = female substance female*substance / ss3;
run;

The GLM Procedure

Dependent Variable: CESD

Source                     DF    Type III SS    Mean Square  F Value  Pr > F

FEMALE                      1    2463.232928    2463.232928    16.84  <.0001
SUBSTANCE                   2    2540.208432    1270.104216     8.69  0.0002
FEMALE*SUBSTANCE            2     145.924987      72.962494     0.50  0.6075
```

```
> aov1 <- aov(cesd ~ sub * genf, data=ds)
> aov2 <- aov(cesd ~ sub + genf, data=ds)
> resid <- residuals(aov2)
> anova(aov2, aov1)

Analysis of Variance Table

Model 1: cesd ~ sub + genf
Model 2: cesd ~ sub * genf
  Res.Df   RSS  Df Sum of Sq   F Pr(>F)
1    449 65515
2    447 65369   2        146 0.5   0.61
```

```
> options(digits=6)
> logLik(aov1)

'log Lik.' -1768.92 (df=7)

> logLik(aov2)

'log Lik.' -1769.42 (df=5)

> lldiff <- logLik(aov1)[1] - logLik(aov2)[1]
> lldiff

[1] 0.505055

> 1 - pchisq(2*lldiff, 2)

[1] 0.603472

> options(digits=3)
```

```
> summary(aov2)

             Df Sum Sq Mean Sq F value  Pr(>F)
sub           2   2704    1352    9.27 0.00011
genf          1   2569    2569   17.61 3.3e-05
Residuals   449  65515     146
```

There is little evidence (p=0.61) of an interaction, so this term can be dropped. For SAS, this means estimating the reduced model.

```
options ls=74; /* stay in gray box */
ods select overallanova parameterestimates;
proc glm data=k.help;
   class female substance;
   model cesd = female substance / ss3 solution;
run;
```

```
The GLM Procedure

Dependent Variable: CESD

                                        Sum of
Source                      DF         Squares    Mean Square   F Value   Pr > F

Model                        3      5273.13263     1757.71088     12.05   <.0001

Error                      449     65515.35744      145.91394

Corrected Total            452     70788.49007
```

```
The GLM Procedure

Dependent Variable: CESD

                                                  Standard
Parameter                     Estimate               Error    t Value    Pr > |t|

Intercept                  39.13070331 B        1.48571047      26.34      <.0001
FEMALE      0              -5.61922564 B        1.33918653      -4.20      <.0001
FEMALE      1               0.00000000 B         .                .          .
SUBSTANCE alcohol          -0.28148966 B        1.41554315      -0.20      0.8425
SUBSTANCE cocaine          -5.60613722 B        1.46221461      -3.83      0.0001
SUBSTANCE heroin            0.00000000 B         .                .          .
```

The model was already fit in R to allow assessment of the interaction.

```
> aov2

Call:
   aov(formula = cesd ~ sub + genf, data = ds)

Terms:
                 sub  genf Residuals
Sum of Squares  2704  2569     65515
Deg. of Freedom    2     1       449

Residual standard error: 12.1
Estimated effects may be unbalanced
```

If results with the same referent categories used by SAS are desired, the default R design matrix (see 3.1.3) can be changed and the model re-fit.

```
> contrasts(sub) <- contr.SAS(3)
> aov3 <- lm(cesd ~ sub + genf, data=ds)
> summary(aov3)

Call:
lm(formula = cesd ~ sub + genf, data = ds)

Residuals:
   Min     1Q Median     3Q    Max
-32.13  -8.85   1.09   8.48  27.09

Coefficients:
            Estimate Std. Error t value Pr(>|t|)
(Intercept)   39.131      1.486   26.34  < 2e-16
sub1          -0.281      1.416   -0.20  0.84247
sub2          -5.606      1.462   -3.83  0.00014
genfM         -5.619      1.339   -4.20  3.3e-05

Residual standard error: 12.1 on 449 degrees of freedom
Multiple R-squared: 0.0745,      Adjusted R-squared: 0.0683
F-statistic:   12 on 3 and 449 DF,  p-value: 1.35e-07
```

The AIC criteria (3.2.3) can also be used to compare models. In SAS it is available in `proc reg` and `proc mixed`. Here we use `proc mixed`, omitting other output.

```
ods select fitstatistics;
proc mixed data=k.help method=ml;
   class female substance;
   model cesd = female|substance;
run; quit;

The Mixed Procedure

           Fit Statistics

-2 Log Likelihood              3537.8
AIC (smaller is better)        3551.8
AICC (smaller is better)       3552.1
BIC (smaller is better)        3580.6
```

```
ods select fitstatistics;
proc mixed data=k.help method=ml;
   class female substance;
   model cesd = female substance;
run; quit;
ods select all;

The Mixed Procedure

           Fit Statistics

-2 Log Likelihood              3538.8
AIC (smaller is better)        3548.8
AICC (smaller is better)       3549.0
BIC (smaller is better)        3569.4
```

```
> AIC(aov1)

[1] 3552

> AIC(aov2)

[1] 3549
```

The AIC criterion also suggests that the model without the interaction is most appropriate.

3.7.6 Multiple comparisons

We can also carry out multiple comparison (3.3.4) procedures to test each of the pairwise differences between substance abuse groups. In SAS this utilizes the `lsmeans` statement within `proc glm`.

```
ods select diff lsmeandiffcl lsmlines;
proc glm data=k.help;
   class substance;
   model cesd = substance;
   lsmeans substance / pdiff adjust=tukey cl lines;
run; quit;
ods select all;
```

```
The GLM Procedure
Least Squares Means
Adjustment for Multiple Comparisons: Tukey-Kramer

i/j              1              2              3

   1                       0.0009         0.9362
   2        0.0009                        0.0008
   3        0.9362         0.0008
```

```
The GLM Procedure
Least Squares Means
Adjustment for Multiple Comparisons: Tukey-Kramer

      Least Squares Means for Effect SUBSTANCE

            Difference           Simultaneous 95%
               Between        Confidence Limits for
i    j           Means         LSMean(i)-LSMean(j)

1    2        4.951829         1.753296     8.150361
1    3       -0.498086        -3.885335     2.889162
2    3       -5.449915        -8.950037    -1.949793
The GLM Procedure
Least Squares Means
Adjustment for Multiple Comparisons: Tukey-Kramer

Tukey-Kramer Comparison Lines for Least Squares Means of SUBSTANCE

LS-means with the same letter are not significantly different.

          CESD                   LSMEAN
        LSMEAN    SUBSTANCE      Number

A     34.87097    heroin              3
A
A     34.37288    alcohol             1

B     29.42105    cocaine             2
```

The above output demonstrates the results of the `lines` option using the `lsmeans` statement. The letter `A` shown on the left connecting the `heroin` and `alcohol` substances implies that there is not a statistically significant difference between these two groups. Since the `cocaine` substance has the letter `B` and no other group has one, the cocaine group is significantly different from each of the other groups. If instead the `cocaine` and `alcohol` substances **both** had a letter `B` attached, while the `heroin` and `alcohol` substances retained the letter `A` they have in the actual output, only the heroin and cocaine groups would differ significantly, while the alcohol group would differ from neither. This presentation becomes particularly useful as the number of groups increases.

In R, we use the `TukeyHSD()` function.

```
> mult <- TukeyHSD(aov(cesd ~ sub, data=ds), "sub")
> mult

  Tukey multiple comparisons of means
    95% family-wise confidence level
```

```
Fit: aov(formula = cesd ~ sub, data = ds)

$sub
                  diff   lwr   upr p adj
cocaine-alcohol -4.952 -8.15 -1.75 0.001
heroin-alcohol   0.498 -2.89  3.89 0.936
heroin-cocaine   5.450  1.95  8.95 0.001
```

The alcohol group and heroin group both have significantly higher CESD scores than the cocaine group, but the alcohol and heroin groups do not significantly differ from each other (95% CI ranges from -2.8 to 3.8). Figure 3.7.6 provides a graphical display of the pairwise comparisons from R.

```
> plot(mult)
```

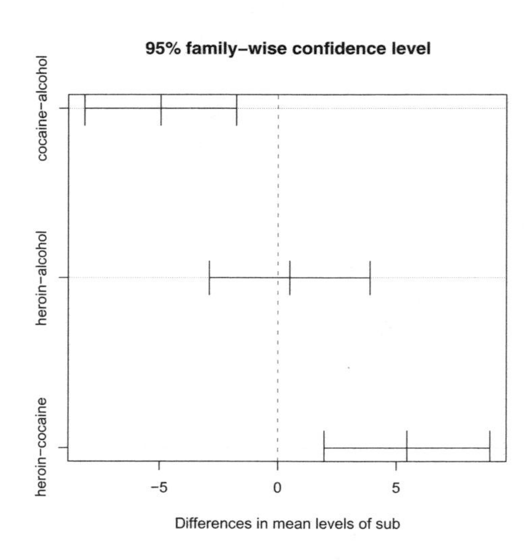

Figure 3.6: Pairwise comparisons

3.7.7 Contrasts

We can also fit contrasts (3.3.3) to test hypotheses involving multiple parameters. In this case, we can compare the CESD scores for the alcohol and heroin groups to the cocaine group. In SAS, to allow checking the contrast, we use the `e` option to the `estimate` statement.

```
ods select contrastcoef estimates;
proc glm data=k.help;
   class female substance;
   model cesd = female substance;
   output out=outanova residual=resid_ch4anova;
   estimate 'A+H = C?' substance 1 -2 1 / e;
run; quit;
ods select all;
```

```
The GLM Procedure

Coefficients for Estimate A+H = C?

                            Row 1

Intercept                       0

FEMALE     0                    0
FEMALE     1                    0

SUBSTANCE alcohol               1
SUBSTANCE cocaine              -2
SUBSTANCE heroin                1
```

```
The GLM Procedure

Dependent Variable: CESD

                                              Standard
Parameter                      Estimate          Error    t Value    Pr > |t|

A+H = C?                     10.9307848     2.42008987       4.52      <.0001
```

```
> library(gmodels)
> fit.contrast(aov2, "sub", c(1,-2,1), conf.int=0.95 )

                 Estimate Std. Error t value Pr(>|t|) lower CI upper CI
sub c=( 1 -2 1 )     10.9       2.42    4.52 8.04e-06     6.17     15.7
```

As expected from the interaction plot (Figure 3.4), there is a statistically significant difference in this one degree of freedom comparison ($p<0.0001$).

Chapter 4

Regression generalizations

This chapter extends the discussion of linear regression introduced in Chapter 3 to include many commonly used statistical models beyond linear regression and ANOVA. Most SAS procedures mentioned in this chapter support the `class` statement for categorical covariates. The CRAN Task View on Statistics for the Social Sciences provides an excellent overview of methods described here and in Chapter 3.

4.1 Generalized linear models

Table 4.1 displays the options for SAS and R to specify link functions and family of distributions for generalized linear models [53]. Description of several specific generalized linear regression models (e.g., logistic and Poisson) can be found in subsequent sections of this chapter.

SAS

```
proc genmod data=ds;
   model y = x1 ... xk / dist=familyname link=linkname;
run;
```

Note: The `class` statement in `proc genmod` is more flexible than that available in many other procedures, notably `proc glm`. However, the default behavior is the same as for `proc glm` (see section 3.1.3).

R

```
glmod1 <- glm(y ~ x1 + ... + xk, family="familyname", link="linkname",
   data=ds)
```

Note: More information on GLM families and links can be found using `help(family)`.

4.1.1 Logistic regression model

HELP example: see 4.6.1

SAS

```
proc logistic data=ds;
   model y = x1 ... xk / or cl;
run;
```

Table 4.1: Generalized linear model distributions supported by SAS and R

Distribution	SAS PROC GENMOD	R glm()
Gaussian	dist=normal	family="gaussian", link="identity", "log" or "inverse"
binomial	dist=binomial	family="binomial", link="logit", "probit", "cauchit", "log" or "cloglog"
gamma	dist=gamma	family="Gamma", link="inverse", "identity" or "log"
Poisson	dist=poisson	family="poisson", link="log", "identity" or "sqrt"
inverse Gaussian	dist=igaussian	family="inverse.gaussian", link="1/mu^2", "inverse", "identity" or "sqrt"
Multinomial	dist=multinomial	See multinom() in nnet library
Negative Binomial	dist=negbin	See negative.binomial() in MASS library
overdispersed	dist=binomial or dist=multinomial with scale=deviance aggregate, or dist=poisson scale=deviance	family="quasi", link="logit", "probit", "cloglog", "identity", "inverse", "log", "1/mu^2" or "sqrt" (see also glm.binomial.disp() in the dispmod library)

Note: For the glm() function in R, the available links for each distribution are listed. The following links are available for all distributions in SAS: identity, log, or power(λ) (where λ is specified by the user). For dichotomous outcomes, complementary log-log (link=cloglog), logit (link=logit), or probit (link=probit) are additionally available. For multinomial distributed outcomes, cumulative complementary log-log (link=cumcll), cumulative logit (link=cumlogit), or cumulative probit (link=cumprobit) are available. Once the family and link functions have been specified, the variance function is implied (with the exception of the quasi family). In SAS, overdispersion is implemented using the scale option to the model statement. To allow overdispersion in Poisson, binomial or multinomial models, use the option scale=deviance; the additional aggregate option is required for the binomial and multinomial. Any valid link listed above may be used.

or

```
proc logistic data=ds;
   model y(event='1') = x1 ... xk;
run;
```

or

```
proc logistic data=ds;
   model r/n = x1 ... xk / or cl; /* events/trials syntax */
run;
```

or

```
proc genmod data=ds;
   model y = x1 ... xk / dist=binomial link=logit;
run;
```

Note: While both procedures will fit logistic regression models, `proc logistic` is likely to be more useful for ordinary logistic regression than `proc genmod`. The former allows options such as those printed above in the first `model` statement, which produce the odds ratios (and their confidence limits) associated with the log-odds estimated by the model. It also produces the area under the ROC curve (the so-called 'c' statistic) by default (see also 5.1.18). Both procedures allow the logit, probit, and complementary log-log links, through the `link` option to the `model` statement; `proc genmod` must be used if other link functions are desired.

The `events/trials` syntax can be used to save storage space for data. In this case, observations with the same covariate values are stored as a single line of data, with the number of observations recorded in one variable (`trials`) and the number with the outcome in another (`events`).

The output from `proc logistic` and `proc genmod` prominently display the level of `y` that is being predicted. The `descending` option to the `proc` statement will reverse the order. Alternatively, the `model` statement in `proc logistic` allows you to specify the target level as shown in the second set of code.

The `class` statement in `proc genmod` is more flexible than that available in many other procedures, notably `proc glm`. Importantly, the default behavior is different than in `proc glm` (see section 3.1.3).

R

```
glm(y ~ x1 + ... + xk, binomial, data=ds)
```

or

```
library(Design)
lrm(y ~ x1 + ... + xk, data=ds)
```

Note: The `lrm()` function within the `Design` package provides the so-called 'c' statistic (area under ROC curve, see also 5.1.18) and the Nagelkerke pseudo-R^2 index [58].

4.1.2 Exact logistic regression

SAS

```
proc logistic data=ds;
   model y = x1 ... xk;
   exact intercept x1;
run;
```

Note: An exact test is generated for each variable listed in the `exact` statement, including if desired the intercept, as shown above. Not all covariates in the `model` statement need be included in the `exact` statement, but all covariates in the `exact` statement must be included in the `model` statement.

R

```
library(elrm)
elrmres <- elrm(y ~ x1 + ... + xk, iter=10000, burnIn=10000, data=ds)
```

Note: The `elrm()` function implements a modified MCMC algorithm to approximate exact conditional inference for logistic regression models [103].

4.1.3 Poisson model

See also 4.1.4 (zero-inflated Poisson)

HELP example: see 4.6.2

SAS

```
proc genmod data=ds;
   model y = x1 ... xk / dist=poisson;
run;
```

Note: The default output from `proc genmod` includes useful methods to assess fit.

R

```
glm(y ~ x1 + ... + xk, poisson, data=ds)
```

Note: It is always important to check assumptions for models. This is particularly true for Poisson models, which are quite sensitive to model departures [32]. One way to assess the fit of the model is by comparing the observed and expected cell counts, and then calculating Pearson's chi-square statistic. This can be carried out using the `goodfit()` function.

```
library(vcd)
poisfit <- goodfit(x, "poisson")
```

The `goodfit()` function carries out a Pearson's χ^2 test of observed vs. expected counts. Other distributions supported include `binomial` and `nbinomial`. R can also create a hanging rootogram [91] to assess the goodness of fit for count models. If the model fits well, then the bottom of each bar in the rootogram should be near zero.

```
library(vcd)
rootogram(poisfit)
```

4.1.4 Zero-inflated Poisson model

HELP example: see 4.6.3

Zero-inflated Poisson models can be used for count outcomes that generally follow a Poisson distribution but for which there are (many) more observed counts of 0 than would be expected. These data can be seen as deriving from a mixture distribution of a Poisson and a degenerate distribution with point mass at zero (see also 4.1.6, zero-inflated negative binomial).

SAS

```
proc genmod data=ds;
   model y = x1 ... xk / dist=zip;
   zeromodel x2 ... xp;
run;
```

Note: The Poisson rate parameter of the model is specified in the `model` statement, with a default log link and alternate link functions available as described in Table 4.1. The zero-probability is modeled as a logistic regression of the covariates specified in the `zeromodel` statement. Support for zero-inflated Poisson models is also available within `proc countreg`.

R

```
library(pscl)
mod <- zeroinfl(y ~ x1 + ... + xk | x2 + ... + xp, data=ds)
```

Note: The Poisson rate parameter of the model is specified in the usual way with a formula as argument to `zeroinfl()`. The default link is `log`. The zero-probability is modeled as a function of the covariates specified after the '`|`' character. An intercept-only model can be fit by including `1` as the second model. Support for zero-inflated negative binomial and geometric models is available.

4.1.5 Negative binomial model

See also 4.1.6 (zero-inflated negative binomial) *HELP example:* see 4.6.4

SAS

```
proc genmod data=ds;
   model y = x1 ... xk / dist=negbin;
run;
```

R

```
library(MASS)
glm.nb(y ~ x1 + ... + xk, data=ds)
```

4.1.6 Zero-inflated negative binomial model

Zero-inflated negative binomial models can be used for count outcomes that generally follow a negative binomial distribution but for which there are (many) more observed counts of 0 than would be expected. These data can be seen as deriving from a mixture distribution of a negative binomial and a degenerate distribution with point mass at zero (see also zero-inflated Poisson, 4.1.4).

SAS

```
proc countreg data=help2;
   model y = x1 ... xk / dist=zinb;
   zeromodel y ~ x2 ... xp;
run;
```

Note: The negative binomial rate parameter of the model is specified in the `model` statement. The zero-probability is modeled as a function of the covariates specified after the $\sim$ in the `zeromodel` statement.

R

```
library(pscl)
mod <- zeroinfl(y ~ x1 + ... + xk | x2 + ... + xp, dist="negbin", data=ds)
```

Note: The negative binomial rate parameter of the model is specified in the usual way with a formula as argument to `zeroinfl()`. The default link is `log`. The zero-probability is modeled as a function of the covariates specified after the '`|`' character. A single intercept for all observations can be fit by including `1` as the model.

4.1.7 Log-linear model

Loglinear models are a flexible approach to analysis of categorical data [2]. A loglinear model of a three-dimensional contingency table denoted by X_1, X_2, and X_3 might assert that the expected counts depend on a 2 way interaction between the first two variables, but that X_3 is independent of all the others:

$$log(m_{ijk}) = \mu + \lambda_i^{X_1} + \lambda_j^{X_2} + \lambda_{ij}^{X_1,X_2} + \lambda_k^{X_3}$$

SAS

```
proc catmod data=ds;
   weight count;
   model x1*x2*x3 =_response_ / pred;
   loglin x1|x2 x3;
run;
```

Note: The variables listed in the `model` statement above describe the n-way table to be analyzed; the term `_response_` is a required keyword indicating a log-linear model. The `loglin` statement specifies the dependence assumptions. The `weight` statement is optional. If used, the `count` variable should contain the cell counts and can be used if the analysis is based on a summary dataset.

R

```
logres <- loglin(table(x1, x2, x3), margin=list(c(1,2), c(3)), param=TRUE)
pvalue <- 1-pchisq(logres$lrt, logres$df)
```

Note: The `margin` option specifies the dependence assumptions. In addition to the `loglin()` function, the `loglm()` function within the `MASS` library provides an interface for log-linear modeling.

4.1.8 Ordered multinomial model

HELP example: see 4.6.6

SAS

```
proc genmod data=ds;
   model y = x1 ... xk / dist=multinomial;
run;
```

or

```
proc logistic data=ds;
   model y = x1 ... xk / link=cumlogit;
run;
```

Note: The `genmod` procedure utilizes a cumulative logit link when `dist=multinomial`, is specified. This compares each level of the outcome with all lower levels. The model implies the proportional odds assumption. The cumulative probit model is available with the `link=cprobit` option to the `model` statement in `proc genmod`. The `proc logistic` implementation provides a score test for the proportional odds assumption.

R

```
library(MASS)
polr(y ~ x1 + ... + xk, data=ds)
```

Note: The default link is logistic; this can be changed to probit, complementary log-log or Cauchy using the `method` option.

4.1.9 Generalized (nominal outcome) multinomial logit

SAS *HELP example:* see 4.6.7

```
proc logistic data=ds;
   model y = x1 ... xk / link=glogit;
run;
```

Note: Each level is compared to a reference level, which can be chosen using the `ref` option e.g., `model y(ref='0') = x1 / link=glogit`.

R

```
library(VGAM)
mlogit <- vglm(y ~ x1 + ... + xk, family=multinomial(), data=ds)
```

4.1.10 Conditional logistic regression model

SAS

```
proc logistic data=ds;
   strata id;
   model y = x1 ... xk;
run;
```

or

```
proc logistic data=ds;
   strata id;
   model y = x1 ... xk;
   exact intercept x1;
run;
```

Note: The variable `id` identifies strata or matched sets of observations. An exact model can be fit using the `exact` statement with list of covariates to be assessed using an exact test, including the intercept, as shown above.

R

```
library(survival)
cmod <- clogit(y ~ x1 + ... + xk + strata(id), data=ds)
```

Note: The variable `id` identifies strata or matched sets of observations. An exact model is fit by default.

4.2 Models for correlated data

There is extensive support within SAS and R for correlated data regression models, including repeated measures, longitudinal, time series, clustered, and other related methods. Throughout this section we assume that repeated measurements are taken on a subject or cluster denoted by variable `id`.

4.2.1 Linear models with correlated outcomes

SAS *HELP example:* see 4.6.10

```
proc mixed data=ds;
   class id;
   model y = x1 ... xk;
   repeated / type=vartype subject=id;
run;
```

or

```
proc mixed data=ds;
   class id;
   model y = x1 ... xk / outpm=dsname;
   repeated ordervar / type=covtypename subject=id;
run;
```

Note: The `solution` option to the `model` statement can be used to get fixed effects parameter estimates in addition to ANOVA tables. The `repeated ordervar` syntax is used when observations within a cluster are a) ordered (as in repeated measurements) b) the placement in the order affects the covariance structure (as in most structures other than independence and compound symmetry) and c) observations may be missing from the beginning or middle of the order. Predicted values for observations can be found using the `outpm` option to the `model` statement as demonstrated in the second block of code. To add to the `outpm` dataset the outcomes and transformed residuals (scaled by the inverse Cholesky root of the marginal covariance matrix), add the `vciry` option to the `model` statement.

The structure of the covariance matrix of the observations is controlled by the `type` option to the `repeated` statement. As of SAS 9.2, there are 36 available structures. Particularly useful options include `un` (unstructured), `cs` (compound symmetry), and `ar(1)` (first-order autoregressive). The full list is available through the on-line help: Contents; SAS Products; SAS Procedures; MIXED; Syntax; REPEATED.

R

```
glsres <- gls(y ~ x1 + ... + xk,
   correlation=corSymm(form = ~ ordervar | id),
   weights=varIdent(form = ~1 | ordervar), ds)
```

Note: The `gls()` function supports estimation of generalized least squares regression models with arbitrary specification of the variance covariance matrix. In addition to a formula interface for the mean model, the analyst specifies a within group correlation structure as well as a description of the within-group heteroscedasticity structure (using the `weights` option). The statement `ordervar | id` implies that associations are assumed within `id`. Other covariance matrix options are available, see `help(corClasses)`.

4.2.2 Linear mixed models with random intercepts

See also 4.2.3 (random slope models), 4.2.4 (random coefficient models), and 6.1.2 (empirical power calculations)

SAS

```
proc mixed data=ds;
   class id;
   model y = x1 ... xk;
   random int / subject=id;
run;
```

Note: The `solution` option to the `model` statement may be required to get fixed effects parameter estimates in addition to ANOVA tables. The `random` statement describes the design matrix for the random effects. Unlike the fixed effects design matrix, specified as usual with the `model` statement, the random effects design matrix includes a random intercept only if it is specified as above. The predicted random intercepts can be printed with the `solution` option to the `random` statement and saved into a dataset using the `ODS` system, e.g an `ods output solutionr=reffs` statement. Predicted values for observations can be found using the `outp=datasetname` and `outpm=datasetname` options to the `model` statement; the `outp` dataset includes the predicted random effects in the predicted values while the `outpm` predictions include only the fixed effects.

R

```
library(nlme)
lmeint <- lme(fixed= y ~ x1 + ... + xk, random = ~ 1 | id,
   na.action=na.omit, data=ds)
```

Note: Best linear unbiased predictors (BLUP's) of the sum of the fixed effects plus corresponding random effects can be generated using the `coef()` function, random effect estimates using the `random.effects()` function, and the estimated variance covariance matrix of the random effects using `VarCorr()`. Normalized residuals (using a Cholesky decomposition, see pages 238-241 of Fitzmaurice et al [22]) can be generated using the `type="normalized"` option when calling `residuals()` using an NLME option (more information can be found using `help(residuals.lme)`). A plot of the random effects can be created using `plot(lmeint)`.

4.2.3 Linear mixed models with random slopes

HELP example: see 4.6.11

See also 4.2.2 (random intercept models) and 4.2.4 (random coefficient models)

SAS

```
proc mixed data=ds;
   class id;
   model y = time x1 ... xk;
   random int time / subject=id type=covtypename;
run;
```

Note: The `solution` option to the `model` statement can be used to get fixed effects parameter estimates in addition to ANOVA tables. Random effects may be correlated with each other (though not with the residual errors for each observation). The structure of the the covariance matrix of the random effects is controlled by the `type` option to the `random` statement. The option most likely to be useful is `type=un` (unstructured); by default, `proc mixed` uses the variance component (`type=vc`) structure, in which the random effects are uncorrelated with each other. The predicted random effects can be printed with the `solution` option to the `random` statement and saved into a dataset using the `ODS` system, e.g an `ods output solutionr=reffs` statement. Predicted values for observations can be found using the `outp=datasetname` and `outpm=datasetname` options to the `model` statement; the `outp` dataset includes the predicted random effects in the predicted values while the `outpm` predictions include only the fixed effects.

R

```
library(nlme)
lmeslope <- lme(fixed=y ~ time + x1 + ... + xk, random = ~ time | id,
   na.action=na.omit, data=ds)
```

Note: The default covariance for the random effects is unstructured (see `help(reStruct)` for other options). Best linear unbiased predictors (BLUP's) of the sum of the fixed effects plus corresponding random effects can be generated using the `coef()` function, random effect estimates using the `random.effects()` function, and the estimated variance covariance matrix of the random effects using `VarCorr()`. A plot of the random effects can be created using `plot(lmeint)`.

4.2.4 More complex random coefficient models

We can extend the random effects models introduced in 4.2.2 and 4.2.3 to 3 or more subject-specific random parameters (e.g., a quadratic growth curve or spline/"broken stick" model [22]). We use $time_1$ and $time_2$ to refer to 2 generic functions of time.

SAS

```
proc mixed data=ds;
   class id;
   model y = time1 time2 x1 ... xk;
   random int time1 time2 / subject=id type=covtypename;
run;
```

Note: The `solution` option to the `model` statement can be used to get fixed effects parameter estimates in addition to ANOVA tables. Random effects may be correlated with each other, though not with the residual errors for each observation. The structure of the covariance matrix of the random effects is controlled by the `type` option to the `random` statement. The option most likely to be useful is `type=un` (unstructured); by default, `proc mixed` uses the variance component (`type=vc`) structure, in which the random effects are uncorrelated with each other. The predicted random effects can be printed with the `solution` option to the `random` statement and saved into a dataset using the `ODS` system, e.g `ods output solutionr=reffs`. Predicted values for observations can be found using the `outp` and `outpm` options to the `model` statement; the `outp` dataset includes the predicted random effects in the predicted values while the `outpm` predictions include only the fixed effects.

R

```
library(nlme)
lmestick <- lme(fixed= y ~ time1 + time2 + x1 + ... + xk,
   random = ~ time1 time2 | id, data=ds, na.action=na.omit)
```

Note: The default covariance for the random effects is unstructured (see `help(reStruct)` for other options). Best linear unbiased predictors (BLUP's) of the sum of the fixed effects plus corresponding random effects can be generated using the `coef()` function, random effect estimates using the `random.effects()` function, and the estimated variance covariance matrix of the random effects using `VarCorr()`. A plot of the random effects can be created using `plot(lmeint)`.

4.2.5 Multilevel models

Studies with multiple levels of clustering can be fit in SAS and R. In a typical example, a study might include schools (as one level of clustering) and classes within schools (a second level of clustering), with individual students within the classrooms providing a response. Generically, we refer to $level_l$ variables which are identifiers of cluster membership at level l. Random effects at different levels are assumed to be uncorrelated with each other.

SAS

```
proc mixed data=ds;
   class id;
   model y = x1 ... xk;
   random int / subject=level1;
   random int / subject=level2;
run;
```

Note: Each `random` statement uses a `subject` option to describe a different clustering structure in the data. There's no theoretical limit to the complexity of the structure or the number of `random` statements, but practical difficulties in fitting the models may be encountered.

R

```
library(nlme)
lmres <- lme(fixed= y ~ x1 + ... + xk, random= ~ 1 | level1 / level2,
   data=ds)
```

Note: A model with k levels of clustering can be fit using the syntax: `level1 / ... / levelk`.

4.2.6 Generalized linear mixed models

HELP example: see 4.6.13 and 6.1.2

SAS

```
proc glimmix data=ds;
   model y = time x1 ... xk / dist=familyname link=linkname;
   random int time / subject=id type=covtypename;
run;
```

Note: Observations sharing a value for `id` are correlated; otherwise, they are assumed independent. Random effects may be correlated with each other (though not with the residual errors for each observation). The structure of the covariance matrix of the random effects is controlled by the `type` option to the `random` statement. There are many available structures. The one most likely to be useful is `un` (unstructured). The full list is available through the on-line help: Contents; SAS Products; SAS Procedures; GLIMMIX; Syntax; RANDOM. As of SAS 9.2, all of the distributions and links shown in Table 4.1 are available, and additionally `dist` can be `beta`, `exponential`, `geometric`, `lognormal`, or `tcentral` (t distribution). An additional link is available for nominal categorical outcomes: `glogit` the generalized logit. Note that the default fitting method relies on an approximation to an integral. The `method=laplace` option to the `proc glimmix` statement will use a numeric integration (this is likely to be time-consuming).

For SAS 9.1 users, `proc glimmix` is available from SAS Institute as a free add-on package: `http://support.sas.com/rnd/app/da/glimmix.html`.

R

```
library(lme4)
glmmres <- lmer(y ~ x1 + ... + xk + (1|id), family=familyval, data=ds)
```

Note: See `help(family)` for details regarding specification of distribution families and link functions.

4.2.7 Generalized estimating equations

HELP example: see 4.6.12

SAS

```
proc genmod data=ds;
   model y = x1 ... xk;
   repeated subject=id / type=corrtypename;
run;
```

Note: The `repeated ordervar` syntax should be used when a) observations within a cluster are a) ordered (as in repeated measurements), b) the placement in the order affects the covariance structure (as in most structures other than independence and compound symmetry), and c) observations may be missing from the beginning or middle of the order. The structure of the working covariance matrix of the observations is controlled by the `type` option to the `repeated` statement. The corrtypes available as of SAS 9.2 include `ar` (first-order autoregressive), `exch` (exchangeable), `ind` (independent), `mdep(m)` (m-dependent), `un` (unstructured), and `user` (a fixed, user-defined correlation matrix).

R

```
library(gee)
geeres <- gee(formula = y ~ x1 + ... + xk, id=id, data=ds,
   family=binomial, corstr="independence")
```

Note: The `gee()` function requires that the dataframe be sorted by subject identifier. Other correlation structures include `"fixed"`, `"stat_M_dep"`, `"non_stat_M_dep"`, `"AR-M"`, and `"unstructured"`. Note that the `"unstructured"` working correlation will only yield correct answers when missing data are monotone, since no ordering options are available in the present release (see `help(gee)` for more information as well as Vincent Carey's "Yet Another GEE solver" `yags` package).

4.2.8 Time series model

Time-series modeling is an extensive area with a specialized language and notation. We make only the briefest approach here. We display fitting an ARIMA (autoregressive integrated moving average) model for the first difference, with first-order auto-regression and moving averages.

In SAS, the procedures to fit time series data are included in the SAS/ETS package. These provide extensive support for time series analysis. However, it is also possible to fit simple auto-regressive models using `proc mixed`. We demonstrate the basic use of `proc arima` (from SAS/ETS). The CRAN Task View on Time Series provides an overview of support available for R.

SAS

```
proc arima data=ds;
   identify var=x(1);
   estimate p=1 q=1;
run;
```

Note: In `proc arima`, the variable to be analyzed is specified in the `identify` statement, with differencing specified in parentheses. The `estimate` statement specifies the order of the autoregression (`p`) and moving average (`q`). Prediction can be accomplished via the `forecast` statement.

R

```
tsobj <- ts(x, frequency=12, start=c(1992, 2))
arres <- arima(tsobj, order=c(1, 1, 1))
```

Note: The `ts()` function creates a time series object, in this case for monthly timeseries data within the variable `x` beginning in February 1992 (the default behavior is that the series starts at time 1 and number of observations per unit of time is 1). The `start` option is either a single number or a vector of two integers which specify a natural time unit and a number of samples into the time unit. The `arima()` function fits an ARIMA model with AR, differencing and MA order all equal to 1.

4.3 Survival analysis

Survival or failure time data, typically consist of the time until the event, as well as an indicator of whether the event was observed or censored at that time. Throughout, we denote the time of measurement with the variable `time` and censoring with a dichotomous variable `cens` = 1 if censored, or = 0 if observed. More information on survival (or failure time, or time-to-event) analysis in R can be found in the CRAN Survival Analysis Task View (see B.6.2). Other entries related to survival analysis include 2.4.4 (log-rank test) and 5.1.19 (Kaplan-Meier plot).

4.3.1 Proportional hazards (Cox) regression model

SAS *HELP example:* see 4.6.14

```
proc phreg data=ds;
   model time*cens(1) = x1 ... xk;
run;
```

Note: SAS supports time varying covariates using programming statements within `proc phreg`. The `class` statement in `proc genmod` is more flexible than that available in many other procedures, notably `proc glm`. However, the default behavior is the same as for `proc glm` (see section 3.1.3).

R

```
library(survival)
coxph(Surv(time, cens) ~ x1 + ... + xk)
```

Note: The Efron estimator is the default; other choices including exact and Breslow can be specified using the `method` option. The `cph()` function within the `Design` package supports time varying covariates, while the `cox.zph()` function within the `survival` package allows testing of the proportionality assumption.

4.3.2 Proportional hazards (Cox) model with frailty

As far as we know, SAS does not incorporate such models as of SAS 9.2. However, an example of accelerated failure time models with frailty can be found in the on-line manual: Contents; SAS Procedures; NLMIXED.

R

```
library(survival)
coxph(Surv(futime, fustat)~ x1 + ... + xk + frailty(id), data=ds)
```

Note: More information on specification of frailty models can be found using `help(frailty)`; support is available for t, Gamma and Gaussian distributions.

4.4 Further generalizations to regression models

4.4.1 Nonlinear least squares model

Nonlinear least squares models [78] can be fit flexibly within SAS and R. As an example, consider the income inequality model described by Sarabia and colleagues [72]:

$$Y = (1 - (1 - X)^p)^{(1/p)}$$

We provide a starting value (0.5) within the interior of the parameter space.

SAS

```
proc nlin data=ds;
   parms p=0.5;
   model y = (1 - ((1-x)**p))**(1/p);
run;
```

R

```
nls(y ~ (1- (1-x)^{p})^(1/{p}), start=list(p=0.5), trace=TRUE)
```

Note: Finding solutions for nonlinear least squares problems is often challenging, see `help(nls)` for information on supported algorithms as well as section 1.8.7 (optimization).

4.4.2 Generalized additive model

HELP example: see 4.6.8

SAS

```
proc gam data=ds;
   model y = spline(x1, df) loess(x2) spline2(x3, x4) ...
      param(x5 ... xk);
run;
```

Note: Specification of a spline or lowess term for variable `x1` is given by `spline(x1)` or `loess(x1)`, respectively, while a bivariate spline fit can be included using `spline2(x1, x2)`. The degrees of freedom can be specified as in `spline(x1, df)`, following a comma in the variable function description, or estimated from the model using generalized cross-validation by including the `method=gcv` option in the model statement. If neither is specified, the default degrees of freedom of 4 is used. Any variables included in `param()` are fit as linear predictors with the usual syntax (3.1.5).

R

```
library(gam)
gam(y ~ s(x1, df) + lo(x2) + lo(x3, x4) + x5 + ... + xk, data=ds)
```

Note: Specification of a smooth term for variable `x1` is given by `s(x1)`, while a univariate or bivariate loess fit can be included using `lo(x1, x2)`. See `gam.s()` and `gam.lo()` within `library(gam)` for details regarding specification of degrees of freedom or span, respectively. Polynomial regression terms can be fit using the `poly()` function.

4.4.3 Robust regression model

Robust regression refers to methods for detecting outliers and/or providing stable estimates when they are present. Outlying variables in the outcome, predictor, or both are considered.

SAS

```
proc robustreg data=ds;
   model y = x1 ... xk / diagnostics leverage;
run;
```

Note: By default, M estimation is performed; other methods are accessed through the `method=method` option to the `proc robustreg` statement, with valid methods including `lts`, `s`, and `mm`.

R

```
library(MASS)
rlm(y ~ x1 + ... + xk, data=ds)
```

Note: The `rlm()` function fits a robust linear model using M estimation. More information can be found in the CRAN Robust Statistical Methods Task View.

4.4.4 Quantile regression model

HELP example: see 4.6.5

Quantile regression predicts changes in the specified quantile of the outcome variable per unit change in the predictor variables; analogous to the change in the mean predicted in least squares regression. If the quantile so predicted is the median, this is equivalent to minimum absolute deviation regression (as compared to least squares regression minimizing the squared deviations).

SAS

```
proc quantreg data=ds;
   model y = x1 ... xk / quantile=0.75;
run;
```

Note: The `quantile` option specifies which quantile is to be estimated (here the 75th percentile). Median regression (i.e., `quantile=0.50`) is performed by default. If multiple quantiles are included (separated by commas) then they are estimated simultaneously, however, standard errors and tests are only carried out when a single quantile is provided.

R

```
library(quantreg)
quantmod <- rq(y ~ x1 + ... + xk, tau=0.75, data=ds)
```

Note: The default for `tau` is 0.5, corresponding to median regression. If a vector is specified, the return value includes a matrix of results.

4.4.5 Ridge regression model

SAS

```
proc reg data=ds ridge=a to b by c;
   model y = x1 ... xk;
run;
```

Note: Each of the values `a, a+c, a + 2c`, ..., `b` is added to the diagonal of the cross-product matrix of $X_1, \ldots, X_k$. Ridge regression estimates are the least squares estimates obtained using this new cross-product matrix.

R

```
library(MASS)
ridgemod <- lm.ridge(y ~ x1 + ... + xk, lambda=seq(from=a, to=b, by=c),
   data=ds)
```

Note: Post-estimation functions supporting `lm.ridge()` objects include `plot()` and `select()`. A vector of ridge constants can be specified using the `lambda` option.

4.5 Further resources

Many of the topics covered in this chapter are active areas of statistical research and many foundational articles are still useful. Here we provide references to texts which serve as accessible references.

Dobson and Barnett [16] is an accessible introduction to generalized linear models, while [53] remains a classic. Agresti [2] describes the analysis of categorical data.

Fitzmaurice, Laird and Ware [22] is an accessible overview of mixed effects methods while [98] reviews these methods for a variety of statistical packages. A comprehensive review of the material in this chapter is incorporated in [20]. The text by Hardin and Hilbe [28] provides a review of generalized estimating equations. The CRAN Task View on Analysis of Spatial Data provides a summary of tools to read, visualize, and analyze spatial data.

Collett [11] is an accessible introduction to survival analysis.

4.6 HELP examples

To help illustrate the tools presented in this chapter, we apply many of the entries to the HELP data. SAS and R code can be downloaded from `http://www.math.smith.edu/sasr/examples`.

In general, SAS output is lengthier than R `summary()` results. We annotate the full output with named ODS objects for logistic regression (section 4.6.1), provide the bulk of results in some examples, and utilize ODS to reduce the output to a few key elements for the sake of brevity for most entries.

```
libname k "c:/book";

data help;
set k.help;
run;
```

```
> options(digits=3)
> options(show.signif.stars=FALSE)
> load("savedfile")
> attach(ds)
```

4.6.1 Logistic regression

In this example, we fit a logistic regression (4.1.1) where we model the probability of being homeless (spending one or more nights in a shelter or on the street in the past six months) as a function of predictors.

We can specify the `param` option to make the SAS reference category match the default in R (see 3.1.3).

```
options ls=74;  /* keep output in grey box */
proc logistic data=help descending;
   class substance (param=ref ref='alcohol');
   model homeless = female i1 substance sexrisk indtot;
run;
```

SAS produces a large number of distinct pieces of output by default. Here we reproduce the ODS name of each piece of output, by running `ods trace on / listing` before the procedure, as introduced in A.7. Each ODS object can also be saved as a SAS dataset using these names with the `ods output` statement as on page 150.

First, SAS reports basic information about the model and the data in the ODS `modelinfo` output.

```
The LOGISTIC Procedure

            Model Information

Data Set                          WORK.HELP
Response Variable                 HOMELESS
Number of Response Levels         2
Model                             binary logit
Optimization Technique            Fisher's scoring
```

Then SAS reports the number of observations read and used, in the ODS `nobs` output. Note that missing data will cause these numbers to differ. Subsetting with the `where` statement (A.6.3) will cause the number of observations displayed here to differ from the number in the dataset.

```
Number of Observations Read            453
Number of Observations Used            453
```

The ODS `responseprofile` output tabulates the number of observations with each outcome, and, importantly, reports which level is being modeled as the event.

```
          Response Profile

 Ordered                      Total
   Value     HOMELESS     Frequency

       1            1           209
       2            0           244

Probability modeled is HOMELESS=1.
```

The ODS `classlevelinfo` output shows the coding for each `class` variable.

```
      Class Level Information

                               Design
Class          Value          Variables

SUBSTANCE      alcohol         0      0
               cocaine         1      0
               heroin          0      1
```

Whether the model converged is reported in the ODS `convergencestatus` output.

```
                  Model Convergence Status

        Convergence criterion (GCONV=1E-8) satisfied.
```

AIC and other fit statistics are produced in the ODS `fitstatistics` output.

Model Fit Statistics

Criterion	Intercept Only	Intercept and Covariates
AIC	627.284	590.652
SC	631.400	619.463
-2 Log L	625.284	576.652

Tests reported in the ODS `globaltests` output assess the joint null hypothesis that all parameters except the intercept equal 0.

Testing Global Null Hypothesis: BETA=0

Test	Chi-Square	DF	Pr > ChiSq
Likelihood Ratio	48.6324	6	<.0001
Score	45.6522	6	<.0001
Wald	40.7207	6	<.0001

The ODS `type3` output contains tests for each covariate (including joint tests for `class` variables with 2 or more values) conditional on all other covariates being included in the model.

Type 3 Analysis of Effects

Effect	DF	Wald Chi-Square	Pr > ChiSq
FEMALE	1	1.0831	0.2980
I1	1	7.6866	0.0056
SUBSTANCE	2	4.2560	0.1191
SEXRISK	1	3.4959	0.0615
INDTOT	1	8.2868	0.0040

The ODS `parameterestimates` output shows the maximum likelihood estimates of the parameters, their standard errors, and Wald statistics and tests for the null hypothesis that the parameter value is 0. Note that in this table, as opposed to the previous one, each level (other than the referent) of any `class` variable is reported separately.

Analysis of Maximum Likelihood Estimates

Parameter	DF	Estimate	Standard Error	Wald Chi-Square	Pr > ChiSq
Intercept	1	-2.1319	0.6335	11.3262	0.0008
FEMALE	1	-0.2617	0.2515	1.0831	0.2980
I1	1	0.0175	0.00631	7.6866	0.0056
SUBSTANCE cocaine	1	-0.5033	0.2645	3.6206	0.0571
SUBSTANCE heroin	1	-0.4431	0.2703	2.6877	0.1011
SEXRISK	1	0.0725	0.0388	3.4959	0.0615
INDTOT	1	0.0467	0.0162	8.2868	0.0040

The ODS `oddsratios` output shows the exponentiated parameter estimates and associated confidence limits.

```
                    Odds Ratio Estimates

                                    Point          95% Wald
Effect                           Estimate      Confidence Limits

FEMALE                              0.770       0.470        1.260
I1                                  1.018       1.005        1.030
SUBSTANCE cocaine vs alcohol        0.605       0.360        1.015
SUBSTANCE heroin  vs alcohol        0.642       0.378        1.091
SEXRISK                             1.075       0.997        1.160
INDTOT                              1.048       1.015        1.082
```

The ODS `association` output shows various other statistics, including the area under the ROC curve, denoted "c" by SAS.

```
Association of Predicted Probabilities and Observed Responses

Percent Concordant       67.8    Somers' D    0.360
Percent Discordant       31.8    Gamma         0.361
Percent Tied              0.4    Tau-a         0.179
Pairs                   50996    c             0.680
```

Within R, we use the `glm()` command to fit the logistic regression model.

```
> logres <- glm(homeless ~ female + i1 + substance + sexrisk + indtot,
+     binomial)
> summary(logres)

Call:
glm(formula = homeless ~ female + i1 + substance + sexrisk +
    indtot, family = binomial)

Deviance Residuals:
   Min      1Q  Median      3Q     Max
 -1.75   -1.04   -0.70    1.13    2.03

Coefficients:
                 Estimate Std. Error z value Pr(>|z|)
(Intercept)      -2.13192    0.63347   -3.37  0.00076
female           -0.26170    0.25146   -1.04  0.29800
i1                0.01749    0.00631    2.77  0.00556
substancecocaine -0.50335    0.26453   -1.90  0.05707
substanceheroin  -0.44314    0.27030   -1.64  0.10113
sexrisk           0.07251    0.03878    1.87  0.06152
indtot            0.04669    0.01622    2.88  0.00399

(Dispersion parameter for binomial family taken to be 1)

    Null deviance: 625.28  on 452  degrees of freedom
Residual deviance: 576.65  on 446  degrees of freedom
AIC: 590.7
Number of Fisher Scoring iterations: 4
```

If the parameter estimates are desired as a dataset, ODS can be used in SAS.

```
ods exclude all;
ods output parameterestimates=helplogisticbetas;
proc logistic data=help descending;
   class substance (param=ref ref='alcohol');
   model homeless = female i1 substance sexrisk indtot;
run;

ods exclude none;
options ls=74;
proc print data=helplogisticbetas;
run;
```

```
                  Class                                                    Prob
Obs   Variable    Val0       DF    Estimate      StdErr     WaldChiSq     ChiSq

 1    Intercept               1     -2.1319      0.6335       11.3262    0.0008
 2    FEMALE                  1     -0.2617      0.2515        1.0831    0.2980
 3    I1                      1      0.0175     0.00631        7.6866    0.0056
 4    SUBSTANCE   cocaine     1     -0.5033      0.2645        3.6206    0.0571
 5    SUBSTANCE   heroin      1     -0.4431      0.2703        2.6877    0.1011
 6    SEXRISK                 1      0.0725      0.0388        3.4959    0.0615
 7    INDTOT                  1      0.0467      0.0162        8.2868    0.0040
```

Similar information can be found in the summary() output object in R.

```
> names(summary(logres))

 [1] "call"           "terms"          "family"         "deviance"
 [5] "aic"            "contrasts"      "df.residual"    "null.deviance"
 [9] "df.null"        "iter"           "deviance.resid" "coefficients"
[13] "aliased"        "dispersion"     "df"             "cov.unscaled"
[17] "cov.scaled"

> coeff.like.SAS <- summary(logres)$coefficients
> coeff.like.SAS

                 Estimate Std. Error z value Pr(>|z|)
(Intercept)       -2.1319    0.63347   -3.37 0.000764
female            -0.2617    0.25146   -1.04 0.297998
i1                 0.0175    0.00631    2.77 0.005563
substancecocaine  -0.5033    0.26453   -1.90 0.057068
substanceheroin   -0.4431    0.27030   -1.64 0.101128
sexrisk            0.0725    0.03878    1.87 0.061518
indtot             0.0467    0.01622    2.88 0.003993
```

4.6.2 Poisson regression

In this example we fit a Poisson regression model (4.1.3) for i1, the average number of drinks per day in the 30 days prior to entering the detox center.

Because `proc genmod` lacks an easy way to specify the reference category, the R results have a different intercept and different effects for the abuse groups.

```
options ls=74;
ods exclude modelinfo nobs classlevels convergencestatus;
proc genmod data=help;
   class substance;
   model i1 = female substance age / dist=poisson;
run;
```

```
The GENMOD Procedure

             Criteria For Assessing Goodness Of Fit

Criterion                     DF           Value        Value/DF

Deviance                     448       6713.8986         14.9864
Scaled Deviance              448       6713.8986         14.9864
Pearson Chi-Square           448       7933.2027         17.7080
Scaled Pearson X2            448       7933.2027         17.7080
Log Likelihood                        16385.3197
Full Log Likelihood                   -4207.6544
AIC (smaller is better)                8425.3089
AICC (smaller is better)               8425.4431
BIC (smaller is better)                8445.8883
```

In the following output, the confidence limits for the parameter estimates, which appear by default in SAS, have been removed.

```
           Analysis Of Maximum Likelihood Parameter Estimates

                                     Standard          Wald
Parameter            DF   Estimate      Error    Chi-Square   Pr > ChiSq

Intercept             1     1.7767     0.0582        930.73       <.0001
FEMALE                1    -0.1761     0.0280         39.49       <.0001
SUBSTANCE  alcohol    1     1.1212     0.0339       1092.72       <.0001
SUBSTANCE  cocaine    1     0.3040     0.0381         63.64       <.0001
SUBSTANCE  heroin     0     0.0000     0.0000           .            .
AGE                   1     0.0132     0.0015         82.52       <.0001
Scale                 0     1.0000     0.0000

NOTE: The scale parameter was held fixed.
```

```
> poisres <- glm(i1 ~ female + substance + age, poisson)
> summary(poisres)

Call:
glm(formula = i1 ~ female + substance + age, family = poisson)
```

```
Deviance Residuals:
   Min       1Q   Median       3Q      Max
 -7.57    -3.69    -1.40     1.04    15.99

Coefficients:
                 Estimate Std. Error z value Pr(>|z|)
(Intercept)       2.89785    0.05827   49.73  < 2e-16
female           -0.17605    0.02802   -6.28  3.3e-10
substancecocaine -0.81715    0.02776  -29.43  < 2e-16
substanceheroin  -1.12117    0.03392  -33.06  < 2e-16
age               0.01321    0.00145    9.08  < 2e-16

(Dispersion parameter for poisson family taken to be 1)

    Null deviance: 8898.9  on 452  degrees of freedom
Residual deviance: 6713.9  on 448  degrees of freedom
AIC: 8425

Number of Fisher Scoring iterations: 6
```

It is always important to check assumptions for models. This is particularly true for Poisson models, which are quite sensitive to model departures. There is support within R for a Pearson's χ^2 goodness of fit test.

```
> library(vcd)
> poisfit <- goodfit(e2b, "poisson")
> summary(poisfit)

         Goodness-of-fit test for poisson distribution

                 X^2 df P(> X^2)
Likelihood Ratio 208 10  3.6e-39
```

The results indicate that the fit is poor ($\chi^2_{10} = 208$, $p < 0.0001$); the Poisson model does not appear to be tenable. This is also seen in the SAS output, which produces several assessments of goodness of fit by default. The deviance value per degree of freedom is high (14.99).

4.6.3 Zero-inflated Poisson regression

A zero-inflated Poisson regression model (4.1.4) might fit better.

```
options ls=74;
ods select parameterestimates zeroparameterestimates;
proc genmod data=help;
   class substance;
   model i1 = female substance age / dist=zip;
   zeromodel female;
run;
```

In the following output, the confidence limits for the parameter estimates, which appear by default in SAS, have been removed.

```
The GENMOD Procedure

         Analysis Of Maximum Likelihood Parameter Estimates

                                         Standard         Wald
Parameter               DF   Estimate       Error   Chi-Square   Pr > ChiSq

Intercept                1     2.2970      0.0599      1471.50       <.0001
FEMALE                   1    -0.0680      0.0280         5.89       0.0153
SUBSTANCE   alcohol      1     0.7609      0.0336       512.52       <.0001
SUBSTANCE   cocaine      1     0.0362      0.0381         0.90       0.3427
SUBSTANCE   heroin       0     0.0000      0.0000          .            .
AGE                      1     0.0093      0.0015        39.86       <.0001
Scale                    0     1.0000      0.0000

NOTE: The scale parameter was held fixed.
 Analysis Of Maximum Likelihood Zero Inflation Parameter Estimates

                               Standard         Wald
Parameter      DF    Estimate     Error    Chi-Square    Pr > ChiSq

Intercept       1     -1.9794    0.1646        144.57        <.0001
FEMALE          1      0.8430    0.2791          9.12        0.0025
```

```
> library(pscl)

pscl 1.03          2008-11-24
```

```
> res <- zeroinfl(i1 ~ female + substance + age | female, data=ds)
> res

Call:
zeroinfl(formula = i1 ~ female + substance + age | female, data = ds)


Count model coefficients (poisson with log link):
     (Intercept)             female  substancecocaine   substanceheroin
         3.05781           -0.06797          -0.72466          -0.76086
             age
         0.00927

Zero-inflation model coefficients (binomial with logit link):
(Intercept)       female
     -1.979        0.843
```

Women are more likely to abstain from alcohol than men (p=0.0025), as well as drink less when they drink (p=0.015). Other significant predictors include `substance` and `age`, though model assumptions for count models should always be carefully verified [32].

4.6.4 Negative binomial regression

A negative binomial regression model (4.1.5) might improve on the Poisson.

```
options ls=74;
ods exclude nobs convergencestatus classlevels modelinfo;
proc genmod data=help;
   class substance;
   model i1 = female substance age / dist=negbin;
run;
```

```
The GENMOD Procedure

             Criteria For Assessing Goodness Of Fit

Criterion                     DF           Value        Value/DF

Deviance                     448        539.5954          1.2045
Scaled Deviance              448        539.5954          1.2045
Pearson Chi-Square           448        444.7200          0.9927
Scaled Pearson X2            448        444.7200          0.9927
Log Likelihood                        18884.8073
Full Log Likelihood                   -1708.1668
AIC (smaller is better)                3428.3336
AICC (smaller is better)               3428.5219
BIC (smaller is better)                3453.0290
```

In the following output, the confidence limits for the parameter estimates, which appear by default in SAS, have been removed.

```
           Analysis Of Maximum Likelihood Parameter Estimates

                                     Standard         Wald
Parameter             DF   Estimate     Error   Chi-Square   Pr > ChiSq

Intercept              1     1.8681    0.2735        46.64       <.0001
FEMALE                 1    -0.2689    0.1272         4.47       0.0346
SUBSTANCE   alcohol    1     1.1488    0.1393        68.03       <.0001
SUBSTANCE   cocaine    1     0.3252    0.1400         5.40       0.0202
SUBSTANCE   heroin     0     0.0000    0.0000          .            .
AGE                    1     0.0107    0.0075         2.04       0.1527
Dispersion             1     1.2345    0.0897

NOTE: The negative binomial dispersion parameter was estimated by maximum
      likelihood.
```

```
> library(MASS)
> nbres <- glm.nb(i1 ~ female + substance + age)
> summary(nbres)
```

```
Call:
glm.nb(formula = i1 ~ female + substance + age,
      init.theta = 0.810015138972117, link = log)

Deviance Residuals:
   Min       1Q  Median       3Q      Max
-2.414  -1.032  -0.278   0.241   2.808

Coefficients:
                 Estimate Std. Error z value Pr(>|z|)
(Intercept)       3.01693    0.28928   10.43  < 2e-16
female           -0.26887    0.12758   -2.11    0.035
substancecocaine -0.82360    0.12904   -6.38  1.7e-10
substanceheroin  -1.14879    0.13882   -8.28  < 2e-16
age               0.01072    0.00725    1.48    0.139

(Dispersion parameter for Negative Binomial(0.81) family taken to be 1)

    Null deviance: 637.82  on 452  degrees of freedom
Residual deviance: 539.60  on 448  degrees of freedom
AIC: 3428

Number of Fisher Scoring iterations: 1


              Theta:  0.8100
          Std. Err.:  0.0589

 2 x log-likelihood:  -3416.3340
```

The Deviance / DF is close to 1, suggesting a reasonable fit. Note that the R and SAS dispersion parameters are inverses of each other.

4.6.5 Quantile regression

In this section, we fit a quantile regression model (4.4.4) of the number of drinks (`i1`) as a function of predictors, modeling the 75th percentile (Q3).

```
ods select parameterestimates;
proc quantreg data=help;
   class substance;
   model i1 = female substance age / quantile=0.75;
run;
```

```
The QUANTREG Procedure

              Parameter Estimates

                                  95% Confidence
Parameter         DF Estimate         Limits

Intercept          1    7.0000  -4.2965  17.2599
FEMALE             1   -2.9091  -7.5765   5.1838
SUBSTANCE alcohol  1   22.6364  18.1744  29.6974
SUBSTANCE cocaine  1    2.5455  -2.6907  10.0198
SUBSTANCE heroin   0    0.0000   0.0000   0.0000
AGE                1    0.1818  -0.2154   0.5752
```

```
> library(quantreg)

Package SparseM (0.79) loaded.  To cite, see citation("SparseM")

> quantres <- rq(i1 ~ female + substance + age, tau=0.75, data=ds)
> summary(quantres)

Call: rq(formula = i1 ~ female + substance + age, tau = 0.75, data = ds)

tau: [1] 0.75

Coefficients:
                 coefficients lower bd upper bd
(Intercept)       29.636       17.274   41.627
female            -2.909       -7.116    3.419
substancecocaine -20.091      -28.348  -15.460
substanceheroin  -22.636      -28.256  -19.115
age                0.182       -0.250    0.521

> detach("package:quantreg")
```

Estimating standard errors and confidence limits is nontrivial in these models, and it is thus unsurprising that the default approaches in R and SAS yield different confidence limits.

Because the `quantreg` package overrides needed functionality in other packages, we `detach()` it after running the `rq()` function (see B.4.5).

4.6.6 Ordinal logit

To demonstrate an ordinal logit analysis (4.6.6), we first create an ordinal categorical variable from the `sexrisk` variable, then model this three level ordinal variable as a function of `cesd` and `pcs`. Note that SAS and R use opposite coding of the reference group for the intercepts (so the estimates are of opposite sign).

```
data help3;
set help;
   sexriskcat = (sexrisk ge 2) + (sexrisk ge 6);
run;
```

```
ods select parameterestimates;
proc logistic data=help3 descending;
   model sexriskcat = cesd pcs;
run;

The LOGISTIC Procedure

              Analysis of Maximum Likelihood Estimates

                                    Standard          Wald
Parameter      DF     Estimate        Error    Chi-Square    Pr > ChiSq

Intercept 2     1      -0.9436       0.5607        2.8326        0.0924
Intercept 1     1       1.6697       0.5664        8.6909        0.0032
CESD            1     -0.00004      0.00759        0.0000        0.9963
PCS             1      0.00521      0.00881        0.3499        0.5542
```

```
> library(MASS)
> sexriskcat <- as.factor(as.numeric(sexrisk>=2) + as.numeric(sexrisk>=6))
> ologit <- polr(sexriskcat ~ cesd + pcs)
> summary(ologit)

Call:
polr(formula = sexriskcat ~ cesd + pcs)

Coefficients:
         Value Std. Error  t value
cesd -3.72e-05    0.00762 -0.00489
pcs   5.23e-03    0.00876  0.59648

Intercepts:
    Value  Std. Error t value
0|1 -1.669  0.562     -2.971
1|2  0.944  0.556      1.698

Residual Deviance: 871.76
AIC: 879.76
```

4.6.7 Multinomial logit

We can fit a multinomial logit (4.6.7) model for the categorized `sexrisk` variable.

```
options ls=74;  /* keep output in grey box */
ods select responseprofile parameterestimates;
proc logistic data=help3 descending;
   model sexriskcat = cesd pcs / link=glogit;
run;
```

```
The LOGISTIC Procedure

          Response Profile

 Ordered                        Total
   Value      sexriskcat     Frequency

       1               2           151
       2               1           244
       3               0            58

Logits modeled use sexriskcat=0 as the reference category.
```

```
               Analysis of Maximum Likelihood Estimates

                                         Standard          Wald
Parameter  sexriskcat  DF  Estimate         Error  Chi-Square  Pr > ChiSq

Intercept  2            1    0.6863        0.9477      0.5244      0.4690
Intercept  1            1    1.4775        0.8943      2.7292      0.0985
CESD       2            1  -0.00672        0.0132      0.2610      0.6095
CESD       1            1   -0.0133        0.0125      1.1429      0.2850
PCS        2            1    0.0105        0.0149      0.4983      0.4802
PCS        1            1   0.00851        0.0140      0.3670      0.5446
```

```
> library(VGAM)
> mlogit <- vglm(sexriskcat ~ cesd + pcs, family=multinomial())
> summary(mlogit)

Call:
vglm(formula = sexriskcat ~ cesd + pcs, family = multinomial())
```

```
Pearson Residuals:
                      Min   1Q Median   3Q Max
log(mu[,1]/mu[,3]) -0.8 -0.7   -0.2 -0.1   3
log(mu[,2]/mu[,3]) -1.3 -1.2    0.8  0.9   1

Coefficients:
               Value Std. Error t value
(Intercept):1 -0.686      0.948    -0.7
(Intercept):2  0.791      0.639     1.2
cesd:1         0.007      0.013     0.5
cesd:2        -0.007      0.009    -0.8
pcs:1         -0.010      0.015    -0.7
pcs:2         -0.002      0.010    -0.2

Number of linear predictors:  2

Names of linear predictors: log(mu[,1]/mu[,3]), log(mu[,2]/mu[,3])

Dispersion Parameter for multinomial family:   1
```

```
Residual Deviance: 870 on 900 degrees of freedom

Log-likelihood: -435 on 900 degrees of freedom

Number of Iterations: 4

> detach("package:VGAM")
```

Because the `VGAM` package overrides needed functionality in other packages, we `detach()` it after running the `vglm()` function (see B.4.5).

4.6.8 Generalized additive model

We can fit a generalized additive model (4.4.2), and generate a plot in `proc gam` using `ODS graphics` (Figure 4.1).

```
ods graphics on;
ods select parameterestimates anodev smoothingcomponentplot;
proc gam data=help plots=components(clm);
   class substance;
   model cesd = param(female) loess(pcs) param(substance) / method=gcv;
run;
ods graphics off;

The GAM Procedure
Dependent Variable: CESD
Regression Model Component(s): FEMALE SUBSTANCE
Smoothing Model Component(s): loess(PCS)

                      Regression Model Analysis
                         Parameter Estimates

                        Parameter        Standard
Parameter                Estimate           Error     t Value     Pr > |t|

Intercept                33.06343         1.09076       30.31       <.0001
FEMALE                    4.40741         1.29026        3.42       0.0007
SUBSTANCE alcohol         0.12608         1.36382        0.09       0.9264
SUBSTANCE cocaine        -3.89240         1.40879       -2.76       0.0060
SUBSTANCE heroin                0               .           .            .


                     Smoothing Model Analysis
                       Analysis of Deviance

                                     Sum of
Source                 DF           Squares     Chi-Square     Pr > ChiSq

Loess(PCS)        2.51843       5041.036452        37.2179         <.0001
```

```
> library(gam)
> gamreg<- gam(cesd ~ female + lo(pcs) + substance)
```

```
> summary(gamreg)

Call: gam(formula = cesd ~ female + lo(pcs) + substance)
Deviance Residuals:
   Min      1Q Median      3Q     Max
-29.16  -8.14   0.81   8.23  29.25

(Dispersion Parameter for gaussian family taken to be 135)
```

```
    Null Deviance: 70788 on 452 degrees of freedom
Residual Deviance: 60288 on 445 degrees of freedom
AIC: 3519

Number of Local Scoring Iterations: 2

DF for Terms and F-values for Nonparametric Effects

             Df Npar Df Npar F Pr(F)
(Intercept) 1.0
female      1.0
lo(pcs)     1.0     3.1   3.77 0.010
substance   2.0

> coefficients(gamreg)

     (Intercept)           female          lo(pcs) substancecocaine
          46.524            4.339           -0.277           -3.956
 substanceheroin
          -0.205
```

The estimated smoothing function is displayed in Figure 4.1.

```
> plot(gamreg, terms=c("lo(pcs)"), se=2, lwd=3)
> abline(h=0)
```

4.6.9 Reshaping dataset for longitudinal regression

A wide (multivariate) dataset can be reshaped (1.5.3) into a tall (longitudinal) dataset. Here we create time-varying variables (with a suffix `tv`) as well as keep baseline values (without the suffix).

In SAS, we do this directly with an `output` statement, putting four lines in the `long` dataset for every line in the original dataset.

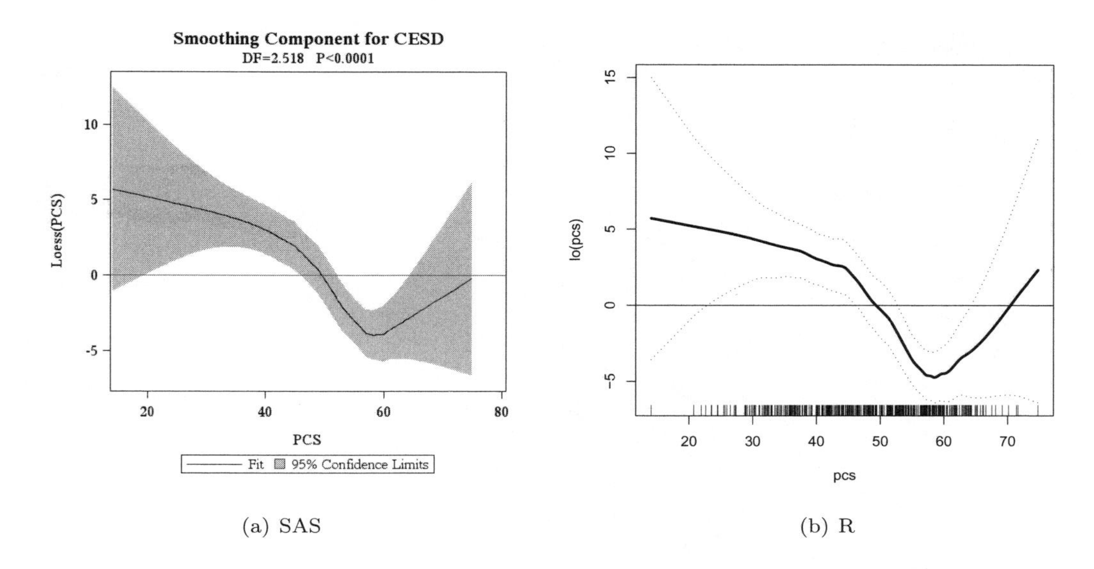

(a) SAS (b) R

Figure 4.1: Scatterplots of smoothed association of PCS with CESD

```
data long;
set help;
   array cesd_a [4] cesd1 - cesd4;
   array mcs_a [4] mcs1 - mcs4;
   array i1_a [4] i11 - i14;
   array g1b_a [4] g1b1 - g1b4;
   do time = 1 to 4;
      cesdtv = cesd_a[time];
      mcstv = mcs_a[time];
      i1tv = i1_a[time];
      g1btv = g1b_a[time];
      output;
      end;
run;
```

In R we use the `reshape()` command.

```
> long <- reshape(ds, idvar="id",
+    varying=list(c("cesd1","cesd2","cesd3","cesd4"),
+    c("mcs1","mcs2","mcs3","mcs4"), c("i11","i12","i13","i14"),
+    c("g1b1","g1b2","g1b3","g1b4")),
+    v.names=c("cesdtv","mcstv","i1tv","g1btv"),
+    timevar="time", times=1:4, direction="long")
> detach(ds)
```

We can check the resulting dataset by printing tables by time. In the code below, we use some options to `proc freq` to reduce the information provided by default.

```
proc freq data=long;
   tables g1btv*time / nocum norow nopercent;
run;

The FREQ Procedure

Table of g1btv by time

g1btv     time

Frequency|
Col Pct  |       1|       2|       3|       4|  Total
---------+--------+--------+--------+--------+
       0 |    219 |    187 |    225 |    245 |    876
         |  89.02 |  89.47 |  91.09 |  92.11 |
---------+--------+--------+--------+--------+
       1 |     27 |     22 |     22 |     21 |     92
         |  10.98 |  10.53 |   8.91 |   7.89 |
---------+--------+--------+--------+--------+
Total         246      209      247      266      968

Frequency Missing = 844
```

```
> table(long$g1btv, long$time)

      1   2   3   4
  0 219 187 225 245
  1  27  22  22  21
```

or by looking at the observations over time for a given individual:

```
proc print data=long;
   where id eq 1;
   var id time cesd cesdtv;
run;

 Obs     ID     time     CESD     cesdtv

 709      1       1       49         7
 710      1       2       49         .
 711      1       3       49         8
 712      1       4       49         5
```

```
> attach(long)
> long[id==1, c("id", "time", "cesd", "cesdtv")]

    id time cesd cesdtv
1.1  1    1   49      7
1.2  1    2   49     NA
1.3  1    3   49      8
1.4  1    4   49      5

> detach(long)
```

This process can be reversed, creating a wide dataset from a tall one.

In SAS, we begin by using `proc transpose` to make a row for each variable with the four time points in it.

```
proc transpose data=long out=wide1 prefix=time;
by notsorted id;
   var cesdtv mcstv i1tv g1btv;
   id time;
run;
```

Note the `notsorted` option to the `by` statement, which allows us to skip an unneeded `proc sort` step and can be used because we know that all the observations for each `id` are stored adjacent to one another.

This results in the following data.

```
proc print data=wide1 (obs=6);
run;

 Obs    ID     _NAME_      time1      time2      time3      time4

   1     2     cesdtv     11.0000       .          .          .
   2     2     mcstv      41.7270       .          .          .
   3     2     i1tv        8.0000       .          .          .
   4     2     g1btv       0.0000       .          .          .
   5     8     cesdtv     18.0000       .       25.0000       .
   6     8     mcstv      36.0636       .       40.6260       .
```

To put the data for each variable onto one line, we merge the data with itself, taking the lines separately and renaming them along the way using the `where` and `rename` data set options (A.6.1).

```
data wide (drop=_name_);
merge
wide1 (where = (_name_="cesdtv")
   rename = (time1=cesd1 time2=cesd2 time3=cesd3 time4=cesd4))
wide1 (where = (_name_="mcstv")
   rename = (time1=mcs1 time2=mcs2 time3=mcs3 time4=mcs4))
wide1 (where = (_name_="i1tv")
   rename = (time1=i11 time2=i12 time3=i13 time4=i14))
wide1 (where = (_name_="g1btv")
   rename = (time1=g1b1 time2=g1b2 time3=g1b3 time4=g1b4));
run;
```

The `merge` without a `by` statement simply places the data from sequential lines in each merged dataset next to each other in the new dataset. Since, here, they are different lines from the same dataset, we know that this is correct. In general, the ability to merge without a by variable in SAS can cause unintended consequences.

The final dataset is as desired.

```
proc print data=wide (obs=2);
   var id cesd1 - cesd4;
run;

Obs    ID    cesd1    cesd2    cesd3    cesd4

  1     2     11        .        .        .
  2     8     18        .       25        .
```

This is a cumbersome process, but is more straightforward than a pure data step approach.

In contrast, converting to a wide format in R can be done with another call to `reshape()`.

```
> wide <- reshape(long,
+    v.names=c("cesdtv", "mcstv", "i1tv", "g1btv"),
+    idvar="id", timevar="time", direction="wide")
> wide[c(2,8), c("id", "cesd", "cesdtv.1", "cesdtv.2", "cesdtv.3",
+    "cesdtv.4")]

    id cesd cesdtv.1 cesdtv.2 cesdtv.3 cesdtv.4
2.1  2   30       11       NA       NA       NA
8.1  8   32       18       NA       25       NA
```

4.6.10 Linear model for correlated data

Here we fit a general linear model for correlated data (modeling the covariance matrix directly, 4.2.1).

```
ods select rcorr covparms solutionf tests3;
proc mixed data=long;
   class time;
   model cesdtv = treat time / solution;
   repeated time / subject=id type=un rcorr=7;
run;
```

In this example, the estimated correlation matrix for the 7th subject is printed (this subject was selected because all four time points were observed).

```
The Mixed Procedure

    Estimated R Correlation Matrix for Subject 7

  Row         Col1         Col2         Col3         Col4

    1       1.0000       0.5843       0.6386       0.4737
    2       0.5843       1.0000       0.7430       0.5851
    3       0.6386       0.7430       1.0000       0.7347
    4       0.4737       0.5851       0.7347       1.0000
```

The estimated elements of the variance-covariance matrix are printed row-wise.

```
Covariance Parameter Estimates

Cov Parm     Subject     Estimate

UN(1,1)      ID            207.21
UN(2,1)      ID            125.11
UN(2,2)      ID            221.29
UN(3,1)      ID            131.74
UN(3,2)      ID            158.39
UN(3,3)      ID            205.36
UN(4,1)      ID           97.8055
UN(4,2)      ID            124.85
UN(4,3)      ID            151.03
UN(4,4)      ID            205.75
```

```
                      Solution for Fixed Effects

                                    Standard
Effect        time    Estimate         Error      DF    t Value    Pr > |t|

Intercept              21.2439        1.0709     381      19.84      <.0001
TREAT                  -0.4795        1.3196     381      -0.36      0.7165
time          1         2.4140        0.9587     381       2.52      0.0122
time          2         2.6973        0.9150     381       2.95      0.0034
time          3         1.7545        0.6963     381       2.52      0.0121
time          4              0             .       .          .           .
```

```
        Type 3 Tests of Fixed Effects

              Num      Den
Effect         DF       DF    F Value    Pr > F

TREAT           1      381       0.13    0.7165
time            3      381       3.53    0.0150
```

```
> library(nlme)
> glsres <- gls(cesdtv ~ treat + as.factor(time),
+    correlation=corSymm(form = ~ time | id),
+    weights=varIdent(form = ~ 1 | time), long, na.action=na.omit)
```

```
> summary(glsres)

Generalized least squares fit by REML
  Model: cesdtv ~ treat + as.factor(time)
  Data: long
   AIC  BIC logLik
  7550 7623  -3760

Correlation Structure: General
 Formula: ~time | id
 Parameter estimate(s):
 Correlation:
  1     2     3
2 0.584
3 0.639 0.743
4 0.474 0.585 0.735
Variance function:
 Structure: Different standard deviations per stratum
 Formula: ~1 | time
 Parameter estimates:
    1     3     4     2
1.000 0.996 0.996 1.033
```

```
Coefficients:
                    Value Std.Error t-value p-value
(Intercept)         23.66     1.098   21.55   0.000
treat               -0.48     1.320   -0.36   0.716
as.factor(time)2     0.28     0.941    0.30   0.763
as.factor(time)3    -0.66     0.841   -0.78   0.433
as.factor(time)4    -2.41     0.959   -2.52   0.012

 Correlation:
                 (Intr) treat  as.()2 as.()3
treat            -0.627
as.factor(time)2 -0.395  0.016
as.factor(time)3 -0.433  0.014  0.630
as.factor(time)4 -0.464  0.002  0.536  0.708

Standardized residuals:
   Min     Q1    Med     Q3    Max
-1.643 -0.874 -0.115  0.708  2.582

Residual standard error: 14.4
Degrees of freedom: 969 total; 964 residual
```

```
> anova(glsres)

Denom. DF: 964
                numDF F-value p-value
(Intercept)         1    1168  <.0001
treat               1       0  0.6887
as.factor(time)     3       4  0.0145
```

A set of parallel boxplots (5.1.7) by time can be generated using the following commands. Results are displayed in Figure 4.2.

```
proc sgpanel data=long;
   panelby time / columns=4;
   vbox cesdtv / category=treat ;
run;
```

```
> library(lattice)
> bwplot(cesdtv ~ as.factor(treat)| time, xlab="TREAT",
+    strip=strip.custom(strip.names=TRUE, strip.levels=TRUE),
+    ylab="CESD", layout=c(4,1), col="black", data=long,
+    par.settings=list(box.rectangle=list(col="black"),
+       box.dot=list(col="black"), box.umbrella=list(col="black")))
```

4.6.11 Linear mixed (random slope) model

Here we fix a mixed effects, or random slope model (4.2.3). Note that in SAS a given variable can be either a `class` variable or not, within a procedure. In this example, we specify a categorical fixed effect of time but a random slope across time treated continuously. We do this by making a copy of the time variable in a new dataset. We save the estimated random effects for later examination, but use `ODS` to suppress their printing.

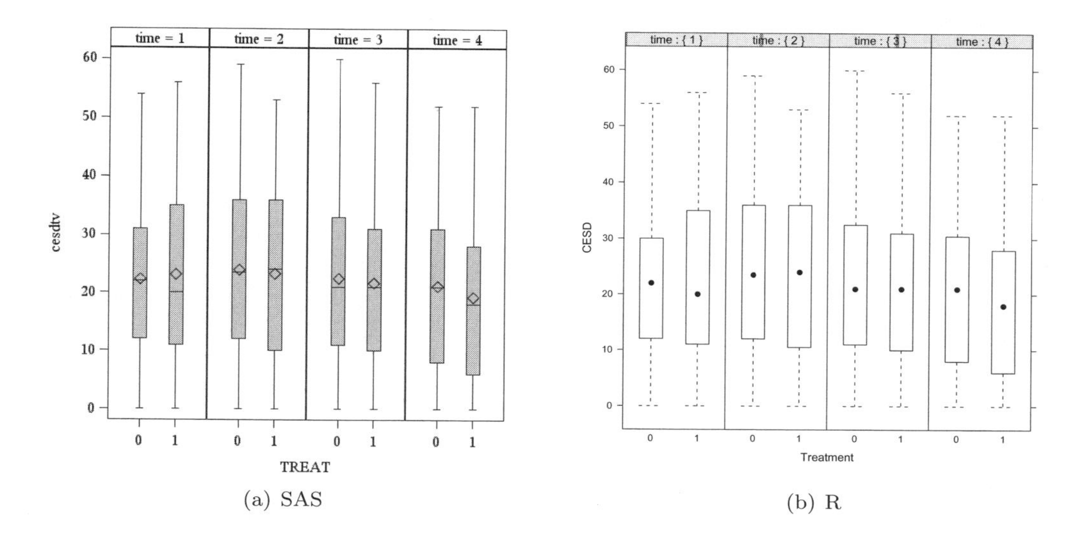

(a) SAS

(b) R

Figure 4.2: Side-by-side box plots of CESD by treatment and time

```
data long2;
set long;
   timecopy=time;
run;
```

To make the first time point the referent, as the R `lme()` function does by default, we first sort by time; then we use the `order=data` option to the `proc mixed` statement.

```
proc sort data= long2; by id descending time; run;

options ls=74;
ods output solutionr=reffs;
ods exclude modelinfo classlevels convergencestatus fitstatistics lrt
      dimensions nobs iterhistory solutionr;
proc mixed data=long2 order=data;
   class timecopy;
   model cesdtv = treat timecopy / solution;
   random int time / subject=id type=un vcorr=20 solution;
run;
```

```
The Mixed Procedure

    Estimated V Correlation Matrix for Subject 20

 Row         Col1         Col2         Col3         Col4

   1       1.0000       0.6878       0.6210       0.5293
   2       0.6878       1.0000       0.6694       0.6166
   3       0.6210       0.6694       1.0000       0.6813
   4       0.5293       0.6166       0.6813       1.0000
```

```
The Mixed Procedure

 Covariance Parameter Estimates

Cov Parm      Subject     Estimate

UN(1,1)       ID            188.43
UN(2,1)       ID          -21.8938
UN(2,2)       ID            9.1731
Residual                   61.5856
```

```
The Mixed Procedure

                      Solution for Fixed Effects

                                         Standard
Effect        timecopy    Estimate          Error      DF    t Value    Pr > |t|

Intercept                  23.8843         1.1066     381      21.58      <.0001
TREAT                      -0.4353         1.3333     292      -0.33      0.7443
timecopy      4            -2.5776         0.9438     292      -2.73      0.0067
timecopy      3            -1.0142         0.8689     292      -1.17      0.2441
timecopy      2           -0.06144         0.8371     292      -0.07      0.9415
timecopy      1                  0              .       .          .           .
```

```
The Mixed Procedure

        Type 3 Tests of Fixed Effects

              Num        Den
Effect         DF         DF     F Value     Pr > F

TREAT           1        292        0.11     0.7443
timecopy        3        292        3.35     0.0195
```

To examine the predicted random effects, or BLUPs, we can look at the `reffs` dataset created by the ODS `output` statement and the `solution` option to the `random` statement. This dataset includes a `subject` variable created by SAS from the `subject` option in the `random` statement. It contains the same information as the `id` variable, but is encoded as a character variable and has some blank spaces in it. In order to easily print the predicted random effects for the subject with `id=1`, we condition using the `where` statement (A.6.3), removing the blanks using the `strip` function (1.4.7).

```
proc print data=reffs;
   where strip(subject) eq '1';
run;
```

```
                                              StdErr
Obs   Effect         Subject    Estimate        Pred      DF     tValue     Probt

  1   Intercept         1       -13.4805      7.4764     292      -1.80    0.0724
  2   time              1       -0.02392      2.3267     292      -0.01    0.9918
```

We can check the predicted values for an individual (incorporating their predicted random effect) using the `outp` option as well as the marginal predicted mean from the `outpm` option to the `model` statement. Here we suppress all output, then print the observed and predicted values for one subject.

```
ods exclude all;
proc mixed data=long2 order=data;
   class timecopy;
   model cesdtv = treat timecopy / outp=lmmp outpm=lmmpm;
   random int time / subject=id type=un;
run;
ods select all;
```

The `outp` dataset has the predicted mean conditional on each subject. The `outpm` dataset has the marginal means. If we want to see them in the same dataset, we can merge them (1.5.7). Note that because the input dataset (`long2`) used in `proc mixed` was sorted, the output datasets are also sorted. Otherwise, a `proc sort` step would be needed for each dataset to be merged. Since both the datasets contain a variable `pred`, we rename one of the variables as we merge the datasets.

```
data lmmout;
merge lmmp lmmpm (rename = (pred=margpred));
   by id descending time;
run;

proc print data=lmmout;
   where id eq 1;
   var id time cesdtv pred margpred;
run;

 Obs    ID    time    cesdtv      Pred       margpred

   1     1      4        5      7.29524      20.8714
   2     1      3        8      8.88264      22.4349
   3     1      2        .      9.85929      23.3876
   4     1      1        7      9.94464      23.4490
```

In R we mimic this process by creating an `as.factor()` version of time. As an alternative, we could nest the call to `as.factor()` within the call to `lme()`.

```
> attach(long)
> tf <- as.factor(time)
> library(nlme)
> lmeslope <- lme(fixed=cesdtv ~ treat + tf,
+    random=~ time |id, data=long, na.action=na.omit)
> print(lmeslope)
```

```
Linear mixed-effects model fit by REML
  Data: long
  Log-restricted-likelihood: -3772
  Fixed: cesdtv ~ treat + tf
(Intercept)        treat          tf2          tf3          tf4
   23.8843      -0.4353      -0.0615      -1.0142      -2.5776

Random effects:
 Formula: ~time | id
 Structure: General positive-definite, Log-Cholesky parametrization
            StdDev Corr
(Intercept) 13.73  (Intr)
time         3.03  -0.527
Residual     7.85

Number of Observations: 969
Number of Groups: 383
```

```
> anova(lmeslope)

            numDF denDF F-value p-value
(Intercept)     1   583    1163  <.0001
treat           1   381       0  0.7257
tf              3   583       3  0.0189
```

In R, we use the `random.effects` and `predict()` functions to find the predicted random effects and predicted values, respectively.

```
> reffs <- random.effects(lmeslope)
> reffs[1,]

  (Intercept)    time
1       -13.5 -0.0239
```

```
> predval <- predict(lmeslope, newdata=long, level=0:1)
> predval[id==1,]

    id predict.fixed predict.id
1.1  1          23.4       9.94
1.2  1          23.4       9.86
1.3  1          22.4       8.88
1.4  1          20.9       7.30
```

```
> vc <- VarCorr(lmeslope)
> summary(vc)

   Variance   StdDev       Corr
   9.17:1    3.03:1         :1
  61.58:1    7.85:1   -0.527:1
 188.43:1   13.73:1   (Intr):1

> detach(long)
```

The `VarCorr()` function calculates the variances, standard deviations, and correlations between the random effects terms, as well as the within-group error variance and standard deviation.

4.6.12 Generalized estimating equations

We fit a GEE model (4.2.7), using an exchangeable working correlation matrix and empirical variance [46]. Note that in the current release of the `gee` package, unstructured working correlations are not supported with nonmonotone missingness. SAS does support this model, using the syntax below (results not shown).

```
proc genmod data=long2 descending;
   class timecopy id;
   model g1btv = treat time / dist=bin;
   repeated subject = id / within=timecopy type=un corrw;
run;
```

To show equivalence between the two systems, we fit the exchangeable correlation structure in SAS as well. In this case the `within` option to the `repeated` statement is not needed.

```
ods select geeemppest geewcorr;
proc genmod data=long2 descending;
   class id;
   model g1btv = treat time / dist=bin;
   repeated subject = id / type=exch corrw;
run;
```

```
The GENMOD Procedure

                Analysis Of GEE Parameter Estimates
                 Empirical Standard Error Estimates

                      Standard   95% Confidence
Parameter Estimate     Error         Limits            Z Pr > |Z|

Intercept   -1.8517   0.2723  -2.3854  -1.3180   -6.80   <.0001
TREAT       -0.0087   0.2683  -0.5347   0.5172   -0.03   0.9740
time        -0.1459   0.0872  -0.3168   0.0250   -1.67   0.0942
```

The `corrw` option requests the working correlation matrix be printed.

```
              Col1          Col2          Col3          Col4

Row1        1.0000        0.2994        0.2994        0.2994
Row2        0.2994        1.0000        0.2994        0.2994
Row3        0.2994        0.2994        1.0000        0.2994
Row4        0.2994        0.2994        0.2994        1.0000
```

```
> library(gee)
> sortlong <- long[order(long$id),]
> attach(sortlong)
> geeres <- gee(formula = g1btv ~ treat + time, id=id, data=sortlong,
+    family=binomial, na.action=na.omit, corstr="exchangeable")

(Intercept)       treat        time
    -1.9649      0.0443     -0.1256

> detach(sortlong)
```

In addition to returning an object with results, the `gee()` function displays the coefficients from a model assuming that all observations are uncorrelated. This is also the default behavior for `proc genmod`, though we have suppressed printing these estimates here.

```
> coef(geeres)

(Intercept)       treat        time
   -1.85169    -0.00874    -0.14593

> sqrt(diag(geeres$robust.variance))

(Intercept)       treat        time
     0.2723      0.2683      0.0872

> geeres$working.correlation

      [,1]  [,2]  [,3]  [,4]
[1,] 1.000 0.299 0.299 0.299
[2,] 0.299 1.000 0.299 0.299
[3,] 0.299 0.299 1.000 0.299
[4,] 0.299 0.299 0.299 1.000
```

4.6.13 Generalized linear mixed model

Here we fit a GLMM (4.2.6), predicting recent suicidal ideation as a function of treatment, depressive symptoms (CESD), and time. Each subject is assumed to have their own random intercept.

```
ods select parameterestimates;
proc glimmix data=long;
   model g1btv = treat cesdtv time / dist=bin solution;
   random int / subject=id;
run;

The GLIMMIX Procedure

                   Solutions for Fixed Effects

                            Standard
Effect          Estimate       Error       DF    t Value     Pr > |t|

Intercept        -4.3572      0.4831      381      -9.02       <.0001
TREAT           -0.00749      0.2821      583      -0.03       0.9788
cesdtv           0.07820     0.01027      583       7.62       <.0001
time            -0.09253      0.1111      583      -0.83       0.4051
```

For many generalized linear mixed models, the likelihood has an awkward shape, and maximizing it can be difficult. In such cases, care should be taken to ensure that results are correct. In such settings, it is useful to use numeric integration, rather than the default approximation used by `proc glimmix`; this can be requested using the `method=laplace` option to the `proc glimmix` statement. When results differ, the maximization based on numeric integration of the actual likelihood should be preferred to the analytic iterative maximization of the approximate likelihood.

```
ods select parameterestimates;
proc glimmix data=long method=laplace;
   model g1btv = treat cesdtv time / dist=bin solution;
   random int / subject=id;
run;

The GLIMMIX Procedure

                    Solutions for Fixed Effects

                              Standard
Effect         Estimate          Error        DF    t Value     Pr > |t|

Intercept       -8.7633         1.2218       381      -7.17       <.0001
TREAT          -0.04164         0.6707       583      -0.06       0.9505
cesdtv           0.1018        0.01927       583       5.28       <.0001
time            -0.2425         0.1730       583      -1.40       0.1616
```

```
> library(lme4)
> glmmres <- lmer(g1btv ~ treat + cesdtv + time + (1|id),
+    family=binomial(link="logit"), data=long)
> summary(glmmres)

Generalized linear mixed model fit by the Laplace approximation
Formula: g1btv ~ treat + cesdtv + time + (1 | id)
   Data: long
 AIC BIC logLik deviance
 480 504   -235      470
Random effects:
 Groups Name        Variance Std.Dev.
 id     (Intercept) 32.6     5.71
Number of obs: 968, groups: id, 383

Fixed effects:
            Estimate Std. Error z value Pr(>|z|)
(Intercept)  -8.7632     1.2802   -6.85  7.6e-12
treat        -0.0417     1.2159   -0.03     0.97
cesdtv        0.1018     0.0237    4.30  1.7e-05
time         -0.2426     0.1837   -1.32     0.19

Correlation of Fixed Effects:
       (Intr) treat  cesdtv
treat  -0.480
cesdtv -0.641 -0.025
time   -0.366  0.009  0.028
```

4.6.14 Cox proportional hazards model

Here we fit a proportional hazards model (4.3.1) for the time to linkage to primary care, with randomization group, age, gender, and CESD as predictors.

```
options ls=74;
ods exclude modelinfo nobs classlevelinfo convergencestatus type3;
proc phreg data=help;
   class treat female;
   model dayslink*linkstatus(0) = treat age female cesd;
run;
```

```
The PHREG Procedure

Summary of the Number of Event and Censored Values

                                        Percent
   Total        Event     Censored     Censored

     431          163          268        62.18
```

```
         Model Fit Statistics

                  Without           With
Criterion      Covariates     Covariates

-2 LOG L         1899.982       1805.368
AIC              1899.982       1813.368
SBC              1899.982       1825.743
```

```
        Testing Global Null Hypothesis: BETA=0

Test                 Chi-Square       DF     Pr > ChiSq

Likelihood Ratio        94.6132        4         <.0001
Score                   92.3599        4         <.0001
Wald                    76.8717        4         <.0001
```

```
                Analysis of Maximum Likelihood Estimates

                    Parameter     Standard                              Hazard
Parameter      DF    Estimate        Error  Chi-Square  Pr > ChiSq       Ratio

TREAT      0    1    -1.65185      0.19324     73.0737      <.0001       0.192
AGE             1     0.02467      0.01032      5.7160      0.0168       1.025
FEMALE     0    1     0.32535      0.20379      2.5489      0.1104       1.385
CESD            1     0.00235      0.00638      0.1363      0.7120       1.002
```

In R we request the Breslow estimator, for compatibility with SAS (the default is the Efron estimator).

```
> library(survival)
> survobj <- coxph(Surv(dayslink, linkstatus) ~ treat + age + female +
+     cesd, method="breslow", data=ds)
> print(survobj)

Call:
coxph(formula = Surv(dayslink, linkstatus) ~ treat + age + female +
    cesd, data = ds, method = "breslow")


          coef exp(coef) se(coef)      z     p
treat   1.65186     5.217  0.19324  8.548 0.000
age     0.02467     1.025  0.01032  2.391 0.017
female -0.32535     0.722  0.20379 -1.597 0.110
cesd    0.00235     1.002  0.00638  0.369 0.710

Likelihood ratio test=94.6  on 4 df, p=0  n=431 (22 observations deleted
    due to missingness)
```

Chapter 5

Graphics

This chapter describes how to create graphical displays, such as scatterplots, boxplots, and histograms. We provide a broad overview of the key ideas and techniques that are available. Additional discussion of ways to annotate displays and change defaults to present publication quality figures is included, as are details regarding how to output graphics in a variety of file formats (section 5.4). Because graphics are useful to visualize analyses, examples appear throughout the HELP sections at the end of most of the chapters of the book.

Producing graphics for data analysis is simple and direct in both programs. Producing graphics for publication is more complex and typically requires a great deal of time to achieve the desired appearance. Our intent is to provide sufficient guidance that most effects can be achieved, but further investigation of the documentation and experimentation will doubtless be necessary for specific needs. There are a huge number of options: we aim to provide a roadmap as well as examples to illustrate the power of both packages.

Base SAS supplies character-based plot procedures, but we focus on procedures to create higher-resolution output using SAS/GRAPH. With version 9.2, SAS adds several powerful ways to generate graphics. One is through statements available in existing procedures, as demonstrated in Figure 2.1 (p. 80). Another is `ods graphics` (A.7.3), as demonstrated for example in Figure 4.1 (p. 161). This approach allows graphical output to be produced easily when generating statistical output. Finally, new procedures are introduced in SAS 9.2 which flexibly generate a variety of graphics especially useful in statistical analysis (for an example, see 5.6.2).

While many graphics in R can be generated using one command, figures are often built up element by element. For example, an empty box can be created with a specific set of x and y axis labels and tick marks, then points can be added with different printing characters. Text annotations can then be added, along with legends and other additional information (see 3.7.1). The Graphics Task View (`http://cran.r-project.org/web/views`) provides a comprehensive listing of functionality to create graphics in R.

As with SAS, a somewhat intimidating set of options is available, some of which can be specified using the `par()` graphics parameters (see section 5.3), while others can be given as options to plotting commands (such as `plot()` or `lines()`).

R provides a number of graphics devices to support different platforms and formats. The default varies by platform (`Windows()` under Windows, `X11()` under Linux and `quartz()` under modern Mac OS X distributions). A device is created automatically when a plotting command is run, or a device can be started in advance to create a file in a particular format (e.g., the `pdf()` device).

A series of powerful add-on packages to create sophisticated graphics are available within R. These include the `grid` package [56], the `lattice` library [74], the `ggplot2` library and the `ROCR` package for receiver operating characteristic curves [80]. Running `example()` for a specified function of interest is particularly helpful for commands shown in this chapter, as is `demo(graphics)`.

5.1 A compendium of useful plots

5.1.1 Scatterplot

HELP example: see 3.7.1

See 5.1.2 (scatterplot with multiple y values) and 5.1.17 (matrix of scatterplots)

SAS

```
proc gplot data=ds;
   plot y*x;
run; quit;
```

or

```
proc sgscatter data=ds;
   plot y*x;
run;
```

Note: The `sgpanel` and `sgplot` procedures in SAS 9.2 also generate scatter plots; `proc sgscatter` is particularly useful for scatterplot matrices (5.1.17).

R

```
plot(x, y)
```

Note: Many objects within R have default plotting functions (e.g., for a linear model object, `plot.lm()` is called). More information can be found using `methods(plot)`. Specifying `type="n"` causes nothing to be plotted (but sets up axes and draws boxes, see 1.13.5). This technique is often useful if a plot is built up part by part.

5.1.2 Scatterplot with multiple y values

HELP example: see 5.6.1

See also 5.1.17 (matrix of scatterplots)

SAS

```
proc gplot data=ds; /* create 1 plot with a single y axis */
   plot (y1 ... yk)*x / overlay;
run; quit;
```

or

```
proc gplot data=ds; /* create 1 plot with 2 separate y axes */
   plot y1*x;
   plot2 y2*x;
run; quit;
```

Note: The first code generates a single graphic with all the different Y values plotted. In this case a simple legend can be added with the `legend` option to the `plot` statement, e.g., `plot (y1 y2)*x / overlay legend`. A fully-controllable legend can be added with a `legend` statement as in Figure 1.1.

The second code generates a single graphic with two y-axes. The scale for Y_1 appears on the left and for Y_2 appears on the right.

In either case, the `symbol` statements (see entries in 5.2) can be used to control the plotted values and add interpolated lines as in 5.6.1. SAS will plot each Y value in a different color and/or symbol by default. The `overlay` option and `plot2` statements are not mutually exclusive, so that several variables can be plotted on each Y axis scale.

Using the statement `plot (y1 ... yk)*x` without the `overlay` option will create k separate plots, identical to k separate `proc gplot` procedures. Adding the `uniform` option to the `proc gplot` statement will create k plots with a common y-axis scale.

R

```
plot(x, y1, pch=pchval1)   # create 1 plot with single y-axis
points(x, y2, pch=pchval2)
...
points(x, yk, pch=pchvalk)
```

or

```
# create 1 plot with 2 separate y axes
addsecondy <- function(x, y, origy, yname="Y2") {
   prevlimits <- range(origy)
   axislimits <- range(y)
   axis(side=4, at=prevlimits[1] + diff(prevlimits)*c(0:5)/5,
      labels=round(axislimits[1] + diff(axislimits)*c(0:5)/5, 1))
   mtext(yname, side=4)
   newy <- (y-axislimits[1])/(diff(axislimits)/diff(prevlimits)) +
      prevlimits[1]
   points(x, newy, pch=2)
}
plottwoy <- function(x, y1, y2, xname="X", y1name="Y1", y2name="Y2")
{
   plot(x, y1, ylab=y1name, xlab=xname)
   addsecondy(x, y2, y1, yname=y2name)
}
plottwoy(x, y1, y2, y1name="Y1", y2name="Y2")
```

Note: To create a figure with a single y axis value, it is straightforward to repeatedly call `points()` or other functions to add elements.

In the second example, two functions `addsecondy()` and `plottwoy()` are defined to add points on a new scale and an appropriate axis on the right. This involves rescaling and labeling the second axis (`side=4`) with 6 tick marks, as well as rescaling the `y2` variable.

5.1.3 Barplot

While not typically an efficient graphical display, there are times when a barplot is appropriate to display counts by groups.

SAS

```
proc gchart data=ds;
   hbar x1 / sumvar=x2 type=mean;
run; quit;
```

or

```
proc sgplot data=ds;
   hbar x1 / response=x2 stat=mean;
run;
```

Note: The above code produces one bar for each level of X_1 with the length determined by the mean of X_2 in each level. Without the `type=mean` or `stat=mean` option, the length would be the sum of x_2 in each level. With no options, the length of each bar is measured in the number of observations in each level of X_1. The `hbar` statement can be replaced by the `vbar` statement (with identical syntax) to make vertical bars, while the `hbar3d` and `vbar3d` (in `proc gchart` only) make bars with a three-dimensional appearance. Options in `proc gchart` allow display of reference lines, display of statistics, grouping by an additional variable, and many other possibilities. The `sgplot` procedure can also produce similar dot plots using the `dot` statement.

R

```
barplot(table(x1, x2), legend=c("grp1", "grp2"), xlab="X2")
```

or

```
library(lattice)
barchart(table(x1, x2, x3))
```

Note: The input for the `barplot()` function is given as the output of a one or two-dimensional contingency table, while the `barchart()` function within `library(lattice)` supports three dimensional tables (see `example(barplot)` and `example(barchart)`). A similar `dotchart()` function produces a horizontal slot for each group with a dot reflecting the frequency.

5.1.4 Histogram

HELP example: see 2.6.1

The example in section 2.6.1 demonstrates how to annotate a histogram with an overlaid normal or kernel density estimates. Similar estimates are available for all other densities supported within R (see Table 1.1) and for the beta, exponential, gamma, lognormal, Weibull, and other densities within SAS.

SAS

```
proc univariate data=ds;
   histogram x1 ... xk;
run;
```

Note: The `sgplot` and `sgpanel` procedures also generate histograms, but allow fewer options.

R

```
hist(x)
```

Note: The default behavior for a histogram is to display frequencies on the vertical axis; probability densities can be displayed using the `freq=FALSE` option. The default title is given by `paste("Histogram of" , x)` where `x` is the name of the variable being plotted; this can be changed with the `main` option.

5.1.5 Stem-and-leaf plot

HELP example: see 3.7.3

Stem-and-leaf plots are text-based graphics that are particularly useful to describe the distribution of small datasets.

SAS

```
proc univariate plot data=ds;
   var x;
run;
```

Note: The stem-and-leaf plot is accompanied by a box plot; the `plot` option also generates a text-based normal Q-Q plot. To produce only these plots, use an `ods select plots` statement before the `proc univariate` statement.

R

```
stem(x)
```

Note: The `scale` option can be used to increase or decrease the number of stems (default value is 1).

5.1.6 Boxplot

See also 5.1.7 (side-by-side boxplots) *HELP example:* see 3.7.5 and 4.6.10

SAS

```
data ds2;
set ds;
   int=1;
run;

proc boxplot data=ds;
   plot x * int;
run;
```

or

```
proc sgplot data=ds;
   vbox x;
run;
```

Note: The `boxplot` procedure is designed to produce side-by-side boxplots (5.1.7). To generate a single boxplot with this procedure, create a variable with the same value for all observations, as above, and make a side-by-side boxplot based on that variable. The `sgplot` procedure also allows the `hbox` statement, which produces a horizontal boxplot.

R

```
boxplot(x)
```

Note: The `boxplot()` function allows sideways orientation using the `horizontal=TRUE` option.

5.1.7 Side-by-side boxplots

See also 5.1.6 (boxplots) *HELP example:* see 3.7.5 and 4.6.10

SAS

```
proc boxplot data=ds;
   plot y * x;
run;
```

or

```
proc boxplot data=ds;
   plot (y1 ... yk) * x (z1 ... zp);
run;
```

or

```
proc sgplot data=ds;
   vbox x / category=y;
run;
```

Note: The first, basic `proc boxplot` code generates a box describing Y for each level of X. The second, more general `proc boxplot` code generates a box for each of $Y_1, Y_2, \ldots, Y_k$ for each level of X, further grouped by $Z_1, Z_2, \ldots, Z_p$. The example in Figure 3.5 demonstrates customization.

The `proc sgplot` code results in boxes of x for each value of y; the similar `hbox` statement makes horizontal boxplots. The `sgpanel` procedure can produce multiple side-by-side boxplots in one graphic using `vbox` or `hbox` statements similar to those shown for `proc sgplot`.

R

```
boxplot(y[x==0], y[x==1], y[x==2], names=c("X=0", "X=1", "X=2"))
```

or

```
boxplot(y ~ x)
```

or

```
library(lattice)
bwplot(y ~ x)
```

Note: The `boxplot()` function can be given multiple arguments of vectors to display, or can use a formula interface (which will generate a boxplot for each level of the variable `x`). A number of useful options are available, including `varwidth` to draw the boxplots with widths proportional to the square root of the number of observations in that group, `horizontal` to reverse the default orientation, `notch` to display notched boxplots, and `names` to specify a vector of labels for the groups. Boxplots can also be created using the `bwplot()` function in `library(lattice)`.

5.1.8 Normal quantile-quantile plot

HELP example: see 3.7.3

Quantile-quantile plots are a commonly used graphical technique to assess whether a univariate sample of random variables is consistent with a Gaussian (normal) distribution.

SAS

```
proc univariate data=ds plot;
   var x;
run;
```

or

```
proc univariate data=ds;
   var x;
   qqplot x;
run;
```

Note: The normal Q-Q plot from the `plot` option is a text-based version; it is accompanied by a stem-and-leaf and a box plot. The plot from the `qqplot` statement is a graphics version. Q-Q plots for other distributions are also available as options to the `qqplot` statement.

R

```
qqnorm(x)
qqline(x)
```

Note: The `qqline()` function adds a straight line which goes through the first and third quartiles.

5.1.9 Interaction plots

HELP example: see 3.7.5

Interaction plots are used to display means by two variables (as in a two-way analysis of variance, 3.1.8).

SAS

```
ods graphics on;
proc glm data=ds;
   class x1 x2;
   model y = x1|x2;
run;
```

Note: In the above, the interaction plot is produced as default output when `ods graphics` are `on` (A.7.3); the `ods select` statement can be used if only the graphic is desired. In addition, an interaction plot can be generated using the `means` and `gplot` procedures (as shown in 3.7.5).

R

```
interaction.plot(x1, x2, y)
```

Note: The default statistic to compute is the mean; other options can be specified using the `fun` option.

5.1.10 Plots for categorical data

A variety of less traditional plots can be used to graphically represent categorical data. While these tend to have a low data to ink ratio, they can be useful in figures with repeated multiples [88]. It is not straightforward to generate these plots in SAS.

R

```
mosaicplot(table(x, y, z))
assocplot(table(x, y))
```

Note: The `mosaicplot()` function provides a graphical representations of a two dimensional or higher contingency table, with the area of each box representing the number of observations in that cell. The `assocplot()` function can be used to display the deviations from independence for a two dimensional contingency table. Positive deviations of observed minus expected counts are above the line and colored black, while negative deviations are below the line and colored red.

5.1.11 Conditioning plot

HELP example: see 5.6.2

A conditioning plot is used to display a scatter plot for each level of one or two classification variables, as below.

SAS

```
proc sgpanel data=ds;
   panelby x2 x3;
   scatter x=x1 y=y;
run;
```

Note: A similar plot can be generated with a boxplot, histogram, or other contents in each cell of $X_2 * X_3$ using other `sgplot` statements in place of the `scatter` statement.

R

```
library(lattice)
coplot(y ~ x1 | x2*x3)
```

Note: The `coplot()` function displays plots of `y` and `x1`, stratified by `x2` and `x3`. All variables may be either numeric or factors.

5.1.12 3-D plots

Perspective or surface plots, needle plots, and contour plots can be used to visualize data in three dimensions. These are particularly useful when a response is observed over a grid of two dimensional values.

SAS

```
proc g3d data=ds;
   scatter x*y=z;
run;

proc g3d data=ds;
   plot x*y=z;
run;

proc gcontour data=ds;
   plot x*y=z;
run;
```

Note: The `scatter` statement produces a needle plot, a 3-D scatterplot with lines drawn from the points down to the $z = 0$ plane to help visualize the third dimension. The `grid` option to the `scatter` statement may help in clarifying the plot, while the needles can be omitted with the `noneedle` option. The `x` and `y` vars must be a grid for the `plot` statement in either the `g3d` (where it produces a surface plot) or the `gcontour` procedure; if not, the `g3grid` procedure can be used to smooth values. The `rotate` and `tilt` options to the `plot` and `scatter` statements will show the plot from a different perspective for the `g3d` procedure.

R

```
persp(x, y, z)
contour(x, y, z)
image(x, y, z)

library(scatterplot3d)
scatterplot3d(x, y, z)
```

Note: The values provided for `x` and `y` must be in ascending order.

5.1.13 Circular plot

Circular plots are used to analyse data that wraps (e.g., directions expressed as angles, time of day on a 24 hour clock) [21, 40]. SAS macros for circular statistics, including a `circplot` macro, are available from Dr. Ulric Lund's webpage at Cal Poly San Luis Obispo.

R

```
library(circular)
plot.circular(x, stack=TRUE, bins=50)
```

5.1.14 Sunflower plot

Sunflower plots [17] are designed to display multiple observations (overplotting) at the same plotting position by adding additional components to the plotting symbol based on how many are at that position. Another approach to this problem involves jittering data (see 5.2.4).

The basic `proc plot` produces text graphics and plots different symbols for overplotting. Sunflower plots within SAS can be generated using a macro written by Michael Friendly at York University.

R

```
sunflowerplot(x, y)
```

5.1.15 Empirical cumulative probability density plot

SAS

```
proc univariate data=ds;
   var x;
   cdfplot x;
run;
```

Note: The empirical density plot offered in `proc univariate` is not smoothed, but theoretical distributions can be superimposed as in the histogram plotted in 2.6.1 and using similar syntax. If a smoothed version is required, it may be necessary to estimate the PDF with `proc kde` and save the output (as shown in 2.6.1), then use it to find the corresponding CDF.

R

```
plot(ecdf(x))
```

5.1.16 Empirical probability density plot

HELP example: see 2.6.4,3.7.3

Density plots are non-parametric estimates of the empirical probability density function.

SAS

```
ods graphics on;
proc kde data=ds;
   univar x1 / plots=(density histdensity);
run;
```

or

```
proc univariate data=ds;
   histogram x / kernel;
run;
```

Note: The `kde` procedure includes kernel density estimation using a normal kernel. The `bivar` statement for `proc kde` will generate a joint empirical density estimate. The bandwidth can be controlled with the `bwm` option and the number of grid points by the `ngrid` option to the `univar` or `bivar` statements. The `proc univariate` code generates a graphic (as in 2.1), but no further details.

R

```
# univariate density
plot(density(x))
```

or

```
library(GenKern)
# bivariate density
op <- KernSur(x, y, na.rm=TRUE)
image(op$xords, op$yords, op$zden, col=gray.colors(100), axes=TRUE,
   xlab="x var", ylab="y var"))
```

Note: The bandwidth for `density()` can be specified using the `bw` and `adjust` options, while the default smoother can be specified using the `kernel` option (possible values include the default gaussian, rectangular, triangular, epanechnikov, biweight, cosine, or optcosine). Bivariate density support is provided with the `GenKern` library. Any of the three-dimensional plotting routines (see 5.1.12) can be used to visualize the results.

5.1.17 Matrix of scatterplots

HELP example: see 5.6.5

SAS

```
proc sgscatter data=ds;
   matrix x1 ... xk;
run;
```

Note: The `diagonal` option to the `matrix` statement allows the diagonal cells to show, for example, histograms with empirical density estimates. A similar effect can be produced with `proc sgpanel` as demonstrated in 5.6.5.

R

```
pairs(data.frame(x1, ..., xk))
```

Note: The `pairs()` function is quite flexible, since it calls user specified functions to determine what to display on the lower triangle, diagonal and upper triangle (see `examples(pairs)` for illustration of its capabilities).

5.1.18 Receiver operating characteristic (ROC) curve

HELP example: see 5.6.4

See also 2.2.2 (diagnostic agreement) and 4.1.1 (logistic regression)

Receiver operating characteristic curves can be used to help determine the optimal cut-score to predict a dichotomous measure. This is particularly useful in assessing diagnostic accuracy in terms of sensitivity (the probability of detecting the disorder if it is present), specificity (the probability that a disorder is not detected if it is not present), and the area under the curve (AUC). The variable `x` represents a predictor (e.g., individual scores) and `y` a dichotomous outcome. There is a close connection between the idea of the ROC curve and goodness of fit for logistic regression, where the latter allows multiple predictors to be used. In SAS, ROC curves are embedded in `proc logistic`; to emulate the functions available in the R `ROCR` library [80], just use a single predictor in SAS `proc logistic`.

SAS

```
ods graphics on;
proc logistic data=ds plots(only)=roc;
   model y = x1 ... xk;
run;
ods graphics off;
```

Note: The `plots(only)` option is used to request only the ROC curve be produced, rather than the default inclusion of several additional plots. The probability cutpoint associated with each point on the ROC curve can be printed using `roc(id=prob)` in place of `roc` above.

R

```
library(ROCR)
pred <- prediction(x, y)
perf <- performance(pred, "tpr", "fpr")
plot(perf)
```

Note: The area under the curve (AUC) can be calculated by specifying `"auc"` as an argument when calling the `performance()` function.

5.1.19 Kaplan–Meier plot

See also 2.4.4 (log-rank test)

HELP example: see 5.6.3

SAS

```
ods graphics on;
ods select survivalplot;
proc lifetest data=ds plots=s;
   time time*status(1);
   strata x;
run;
ods graphics off;
```

or

```
proc lifetest data=ds outsurv=survds;
   time time*status(1);
   strata x;
run;
```

```
symbol1 i=stepj r=kx;
proc gplot data=survds;
   plot survival*survtime = x;
run;
```

Note: The second approach demonstrates how to manually construct the plot without using ods graphics (A.7.3). The survival estimates generated by proc lifetest are saved in a new dataset using the outsurv option to the proc lifetest statement; we suppose there are kx levels of x, the stratification variable.

For the plot, a step-function to connect the points is specified using the i=stepj option to the symbol statement. Finally proc gplot with the a*b=c syntax (5.2.2) is called. In this case, survival*survtime=x will plot lines for each of the kx levels of x. Here, survival and survtime are variable names created by proc lifetest. Note that the r=kx option to the symbol statement is shorthand for typing in the same options for symbol1, symbol2, ..., symbolkx statements; here we repeat them for the kx strata specified in x.

R

```
library(survival)
fit <- survfit(Surv(time, status) ~ as.factor(x), data=ds)
plot(fit, conf.int=FALSE, lty=1:length(unique(x)))
```

Note: The Surv() function is used to combine survival time and status, where time is length of follow-up (interval censored data can be accommodated via an additional parameter) and status=1 indicates an event (e.g., death) while status=0 indicated censoring. A stratified model by each level of the group variable x (see also adding legends, 5.2.14 and different line styles, 5.3.9). More information can be found in the CRAN Survival Task View.

5.2 Adding elements

In R, it is relatively simple to add features to graphs which have been generated by one of the functions discussed in section 5.1. For example adding an arbitrary line to a graphic requires only one function call with the two endpoints as arguments (5.2.1). In SAS, such additions are made using a specially-formatted dataset called an annotate dataset; see section 6.4.2 for an example. These datasets contain certain required variable names and values. Perfecting a graphic for publication can be facilitated by detailed understanding of annotate datasets, a powerful low-level tool. Their use is made somewhat easier by a suite of SAS macros, the annotate macros provided with SAS/GRAPH. To use the macros, you must first enable them in the following way.

```
%annomac;
```

You can then call on the macros to draw a line between two points, or plot a circle, and so forth. You do this by creating an annotate dataset and calling the macros within it.

```
data annods;
   %system(x, y, s);
   ...
run;
```

Here the ellipses refer to additional annotate macros. The system macro is useful in getting the macros to work as desired; it defines how the values of x and y in later annotate macros are interpreted as well as the size of the plotted values. For example, to measure in terms of the graphics output area, use the value 3 for the first two parameters in the system macro. This can be useful for drawing outside the axes. More frequently, we find that using the coordinate system of the plot itself is most convenient; using the value 2 for for each parameter will implement this.

5.2.1 Arbitrary straight line

HELP example: see 3.7.1

SAS

```
%annomac;
data annods;
   %system(2,2,2);
   %line(xvalue_1, yvalue_1, xvalue_2, yvalue_2, colorspec, linetype, .01);
run;

proc gplot data=ds;
   plot x*y / anno=annods;
run; quit;
```

Note: See section 5.2 for an overview of `annotate` datasets. The `line` macro draws a line from (`xvalue_1`, `yvalue_1`) to (`xvalue_2`, `yvalue_2`). The line will have the color (5.3.11) specified by `colorspec` and be solid or dashed (5.3.9) as specified in `linetype`. The final entry specified the width of the line, here quite narrow. Another approach would be to add the endpoint values to the original dataset, then use the `symbol` statement and the `a*b=c` syntax of `proc gplot` (5.2.2).

R

```
plot(x, y)
lines(point1, point2)
```

or

```
abline(intercept, slope)
```

Note: The `lines()` function draws a line between the points specified by `point1` and `point2`, which are each vectors with values for the `x` and `y` axes. The `abline()` function draws a line based on the slope-intercept form. Vertical or horizontal lines can be specified using the `v` or `h` option to `abline()`.

5.2.2 Plot symbols

HELP example: see 2.6.2

SAS

```
symbol1 value=valuename;
symbol1 value='plottext';
symbol1 font=fontname value=plottext;
proc gplot data=ds;
   ...
run;
```

or

```
proc gplot data=ds;
   plot y*x = groupvar;
run; quit;
```

Note: The specific characters plotted in `proc gplot` can be controlled using the `value` option to a preceding `symbol` statement as demonstrated in Figure 2.2. The `valuenames` available include `dot`, `point`, `plus`, `diamond`, and `triangle`. They can also be colored with the `color` option and their size changed with the `height` option. A full list of plot symbols can be found in the on-line help: Contents; SAS Products; SAS/GRAPH; Procedures and Statements; Statements; SYMBOL. The list appears approximately two-thirds of the way through the entry. Additionally, any font character or string can be plotted, if enclosed in quotes as in the second `symbol` statement example, or without the quotes if a `font` option is specified as in the third example.

In the second set of code, a unique plot symbol or color is printed for each value of the variable `group`. If there are many values, for example if `groupvar` is continuous, the results can be confusing.

R

```
plot(x, y, pch=pchval)
```

or

```
points(x, y, string, pch=pchval)
```

or

```
library(lattice)
xyplot(x ~ y, group=factor(groupvar), data=ds)
```

or

```
library(ggplot2)
qplot(x, y, col=factor(groupvar), shape=factor(groupvar), data=ds)
```

Note: The `pch` option requires either a single character or an integer code. Some useful values include `20` (dot), `46` (point), `3` (plus), `5` (diamond), and `2` (triangle) (running `example(points)` will display more possibilities). The size of the plotting symbol can be changed using the `cex` option. The vector function `text()` adds the value in the variable `string` to the plot at the specified location. The examples using `xyplot()` and `qplot()` will also generate scatterplots with different plot symbols for each level of `groupvar`.

5.2.3 Add points to an existing graphic

HELP example: see 3.7.1

See also 5.2.2 (specifying plotting character)

SAS

```
%annomac;
data annods;
   %system(2, 2, 2);
   %circle(xvalue, yvalue, radius);
run;

proc gplot data=ds;
   plot x*y / anno=annods;
run; quit;
```

Note: See section 5.2 for an introduction to `annotate` datasets. The `circle` macro draws a circle with the center at (`xvalue,yvalue`) and with a radius determined by the last parameter. A suitably small radius will plot a point. Another approach is to add a value to the original dataset, then use the `symbol` statement and the `a*b=c` syntax of `proc gplot` (5.2.2).

R

```
plot(x, y)
points(x, y)
```

5.2.4 Jitter

HELP example: see 2.6.2

Jittering is the process of adding a negligible amount of noise to each observed value so that the number of observations sharing a value can be easily discerned. This can be accomplished in a data step within SAS or using the built-in function within R.

SAS

```
data ds;
set newds;
   jitterx = x + ((uniform(0) * .4) - .2);
run;
```

Note: The above code replicates the default behavior of R, assuming `x` has a minimum distance between values of 1.

R

```
jitterx <- jitter(x)
```

Note: The default value for the range of the random uniforms is 40% of the smallest difference between values.

5.2.5 OLS line fit to points

HELP example: see 2.6.1

SAS

```
symbol1 interpol=rl;
proc gplot data=ds;
   plot y*x;
run;
```

or

```
proc sgplot data=ds;
   reg x=x y=y;
run;
```

Note: For `proc gplot`, related interpolations which can be specified in the `symbol` statement are `rq` (quadratic fit) and `rc` (cubic fit). Note also that confidence limits for the mean or for individual predicted values can be plotted by appending `clm` or `cli` after `rx` (see 3.5.5 and 3.5.6). The type of line can be modified as described in 5.3.9. For the `proc sgplot` approach, confidence limits can be requested with the `clm` and/or `cli` options to the `reg` statement; polynomial regression curves can be plotted using the `degree` option. Similar plots can be generated by `proc reg` using `ods graphics` (A.7.3) and by the `sgscatter` and `sgpanel` procedures.

R

```
plot(x, y)
abline(lm(y ~ x))
```

Note: The `abline()` function accepts regression objects with a single predictor as input.

5.2.6 Smoothed line

See also 4.6.8 (generalized additive models) *HELP example:* see 2.6.2

SAS

```
symbol1 interpol=splines;
proc gplot data=ds;
   plot y*x;
run;
```

or

```
ods graphics on;
proc loess data=ds;
   model y = x;
ods graphics off;
```

or

```
ods graphics on;
proc gam data=ds plots=all;
   model y = x;
ods graphics off;
```

or

```
proc sgplot;
   loess x=x y=y;
run;
```

Note: The `spline` interpolation in the `symbol` statement smooths a plot using cubic splines with continuous second derivatives. Other smoothing `interpolation` options include `sm`, which uses a cubic spline which minimizes a linear combination of the sum of squares of the residuals and the integral of the square of the second derivative. In that case, an integer between 0 and 99, appended to the `sm` controls the smoothness. Another option is `interpol=lx`, which uses a Lagrange interpolation of degree `x`, where $x = 1, 3, 5$. For all of these smoothers, using the `s` suffix to the method sorts the data internally. If the data are previously sorted, this is not needed. The `sgplot` procedure also offers penalized B-spline smoothing via the `pbspline` statement; the `sgpanel` procedure also includes these smoothers.

R

```
plot(...)
lines(lowess(x, y))
```

Note: The `f` parameter to `lowess()` can be specified to control the proportion of points which influence the local value (larger values give more smoothness). The `supsmu()` (Friedman's 'super smoother') and `loess()` (local polynomial regression fitting) functions are alternative smoothers.

5.2.7 Normal density

HELP example: see 3.7.3

A normal density plot can be added as an annotation to a histogram or empirical density.

SAS

```
proc sgplot data=ds;
   density x;
run;
```

or

```
proc univariate data=ds;
   histogram x / normal;
run;
```

Note: The sgplot procedure will draw the estimated normal curve without the histogram, as shown. The histogram can be added using the histogram statement; the order of the statements determines which element is plotted on top of the other(s). The univariate procedure allows many more distributional curves to be fit; it will generate copious text output unless that is suppressed with the ods select statement.

R

```
hist(x)
xvals <- seq(from=min(x), to=max(x), length=100)
lines(pnorm(xvals, mean(x), sd(x))
```

5.2.8 Marginal rug plot

HELP example: see 2.6.2

A rug plot displays the marginal distribution on one of the margins of a scatterplot. While this is possible in SAS, using annotate datasets or proc sgrender, it is non-trivial.

R

```
rug(x, side=sideval)
```

Note: The rug() function adds a marginal plot to one of the sides of an existing plot (sideval=1 for bottom (default), 2 for left, 3 for top and 4 for right side).

5.2.9 Titles

HELP example: see 2.6.4

SAS

```
title 'Title text';
```

or

```
title1 "Main title";
title2 "subtitle";
```

Note: The title statement is not limited to graphics, but will also print titles on text output. To prevent any title from appearing after having specified one, use a title statement with no quoted title text. Up to 99 numbered title statements are allowed. For graphic applications, font characteristics can be specified with options to the title statement.

R

```
title(main="main", sub="sub", xlab="xlab", ylab="ylab")
```

Note: The title commands refer to the main title, sub-title, x-axis, and y-axis, respectively. Some plotting commands (e.g., hist()) create titles by default, and the appropriate option within those routines needs to be specified when calling them.

5.2.10 Footnotes

SAS

```
footnote 'footnote text';
```

or

```
footnote1 "Main footnote";
footnote2 "subtitle";
```

Note: The `footnote` statement in SAS is not limited to graphics, but will also print footnotes on text output. To prevent any title from appearing after having specified one, use a `footnote` statement with no quoted footnote text. Up to 10 numbered `footnote` statements are allowed. For graphic applications, font characteristics can be specified with options to the `footnote` statement.

R

```
title(sub="sub")
```

Note: The `sub` option for the `title()` function generates a subtitle.

5.2.11 Text

SAS

HELP example: see 2.6.2, 6.4.2

```
%annomac;
data annods;
   %system(2,2,3);
   %label(xvalue, yvalue, "text", color, angle, rotate, size,
      font, position);
run;

proc gplot data=ds;
   plot x*y / anno=annods;
run; quit;
```

Note: See section 5.2 for an introduction to `annotate` datasets. The `label` macro draws the text provided in `text` (`xvalue`, `yvalue`), though a character variable can also be specified, if the quotes are omitted. The remainder of the parameters which define the text are generally self-explanatory with the exception of `size` which is a numeric value measured in terms of the size of the graphics area, and `position` which specifies the location of the specified point relative to the printed text. A value of 5 centers the text on the specified point. Fonts available include SAS and system fonts; a default typical SAS font is `swiss`. SAS font information can be found in the on-line help: Contents; SAS Products; SAS/GRAPH; Concepts; Fonts.

R

```
text(x, y, labels)
```

Note: Each value of the character vector `labels` is displayed at the specified (X,Y) coordinate. The `adj` option can be used to change text justification to the left, center (default) or right of the coordinate. The `srt` option can be used to rotate text, while `cex` controls the size of the text. The `font` option to `par()` allows specification of plain, bold, italic, or bold italic fonts (see the `family` option to specify the name of a different font family).

5.2.12 Mathematical symbols

HELP example: see 1.13.5

In SAS, mathematical symbols can be plotted using the text plotting method described in 5.2.11, specifying a font containing math symbols. These can be found in the documentation: Contents; SAS Products; SAS/GRAPH; Concepts; Fonts. Useful fonts include the `math` and `greek` fonts. Putting equations with subscripts and superscripts into a plot, or mixing fonts, can be very time-consuming.

R

```
plot(x, y)
text(x, y, expression(mathexpression))
```

Note: The `expression()` argument can be used to embed mathematical expressions and symbols (e.g., $\mu = 0$, $\sigma^2 = 4$) in graphical displays as text, axis labels, legends, or titles. See `help(plotmath)` for more details on the form of `mathexpression` and `example(plotmath)` for examples.

5.2.13 Arrows and shapes

SAS

HELP example: see 2.6.4, 5.6.5

```
%annomac;
data annods;
   %system(2,2,3);
   %arrow(xvalue_1, yvalue_1, xvalue_2, yvalue_2, color,
      linetype, size, angle, font);
   %rect(xvalue_1, yvalue_1, xvalue_2, yvalue_2, color, linetype,
      size);
run;

proc gplot data=ds;
   plot x*y / anno=annods;
run; quit;
```

Note: See section 5.2 for an introduction to `annotate` datasets. The `arrow` macro draws an arrow from (`xvalue_1`, `yvalue_1`) to (`xvalue_2`, `yvalue_2`). The `size` is a numeric value measured in terms of the size of the graphics area. The `rect` macro draws a rectangle with opposite corners at (`xvalue_1`, `yvalue_1`) and (`xvalue_2`, `yvalue_2`). The type of line drawn is determined by the value of `linetype`, as discussed in 5.3.9, and the color is determined by the value of `color` as discussed in 5.3.11.

R

```
arrows(x, y)
rect(xleft, ybottom, xright, ytop)
polygon(x, y)

library(plotrix)
draw.circle(x, y, r)
```

Note: The `arrows`, `rect()` and `polygon()` functions take vectors as arguments and create the appropriate object with vertices specified by each element of those vectors.

5.2.14 Legend

HELP example: see 1.13.5, 2.6.4

SAS

```
legend1 mode=share position=(bottom right inside)
   across=ncols frame label=("Legend Title" h=3) value=("Grp1" "Grp2");

proc gplot data=ds;
   plot y*x=group / legend=legend1;
run;
```

Note: The `legend` statement controls all aspects of how the legend will look and where it will be placed. Legends can be attached to many graphics in a manner similar to that demonstrated here for `proc gplot`. Here we show the most commonly used options. An example of using the `legend` statement can be found in Figure 3.1 (p. 112). The `mode` option determines whether the legend shares the graphic output region with the graphic (shown above); other options reserve space or prevent other plot elements from interfering with the graphic. The `position` option places the legend within the plot area. The `across` option specifies the number of columns in the legend. The `frame` option draws a box around the legend. The `label` option describes the text of the legend title, while the `value` option describes the text printed with legend items. Fuller description of the legend statement is provided in the on-line documentation: Contents; SAS Products; SAS/GRAPH; Procedures and Statements; Statements; LEGEND.

R

```
plot(x, y)
legend(xval, yval, legend=c("Grp1","Grp2"), lty=1:2, col=3:4)
```

Note: The `legend()` command can be used to add a legend at the location `(xval, yval)` to distinguish groups on a display. Line styles (5.3.9) and colors (5.3.11) can be used to distinguish the groups. A vector of legend labels, line types and colors can be specified using the `legend`, `lty`, and `col` options, respectively.

5.2.15 Identifying and locating points

SAS

```
symbol1 pointlabel=("#label");
proc gplot data=ds;
   plot y*x;
run;
quit;
```

```
data newds;
set ds;
   label = 'alt=' || x || "," || y;
run;

ods html;
proc gplot data=newds;
   plot y*x / html=label;
run; quit;
ods html close;
```

Note: The former code will print the values of the variable `label` on the plot. The variable `label` must appear in the dataset used in the `proc gplot` statement. Note that this can result in messy plots, and it is advisable when there are many observations to choose or create a `label` variable with mostly missing values.

The latter code will make the value of X and Y appear when the mouse hovers over a plotted data point, as long as the HTML output destination is used. Any text or variable value can be displayed in place of the value of `label`, which in the above entry specifies the observed values of x and y.

R

```
locator(n)
```

Note: The `locator()` function identifies the position of the cursor when the mouse button is pressed. An optional argument `n` specifies how many values to return. The `identify()` function works in the same fashion, but returns the point closest to the cursor.

5.3 Options and parameters

Many options can be given to plots. In many SAS procedures, these are implemented using `goptions`, `symbol`, `axis`, `legend`, or other statements. Details on these statements can be found in the on-line help: Contents; SAS Products; SAS/GRAPH; Procedures and Statements; Statements.

In R, many options are arguments to `plot()`, `par()`, or other high-level functions. Many of these options are described in the documentation for the `par()` function.

5.3.1 Graph size

SAS

```
goptions hsize=Xin vsize=Yin;
```

or

```
ods graphics width=Xin height=Yin;
```

Note: The size in `goptions` can be specified as above in inches (`in`) or as centimeters (`cm`). The size in `ods graphics` (A.7.3) can also be specified as (`cm`), millimeters (`mm`), standard typesetting dimensions (`em`, `en`), or printer's points (`pt`).

R

```
pdf("filename.pdf", width=Xin, height=Yin)
```

Note: The graph size is specified as an optional argument when starting a graphics device (e.g., pdf(), section 5.4.1), with arguments `Xin` and `Yin` given in inches.

5.3.2 Point and text size

SAS

HELP example: see 3.7.5

```
goptions htext=Xin;
title 'titletext' h=Xin;
axis label = ('labeltext' h=Xin);
axis value = ('valuetext' h=Yin);
```

Note: For many graphics statements which produce text, the `h` option controls the size of the printed characters. The default metric is graphic cells, but absolute values in inches and centimeters can also be used as in the `axis` statements shown. The `htext` option to the `goptions` statement affects all text in graphic output unless changed for a specific graphic element.

R

```
plot(x, y, cex=cexval)
```

Note: The `cex` options specified how much the plotting text and symbols should be magnified relative to the default value of 1 (see `help(par)` for details on how to specify this for axis, labels and titles, e.g., `cex.axis`).

5.3.3 Box around plots

HELP example: see 2.6.4

In SAS, some graphics-generating statements accept a `frame` or a (default) `noframe` option, which will draw or prevent drawing a box around the plot.

R

```
plot(x, y, bty=btyval)
```

Note: Control for the box around the plot can be specified using `btyval`, where if the argument is one of `o` (the default), `l`, `7`, `c`, `u`, or `]`, the resulting box resembles the corresponding character, while a value of `n` suppresses the box.

5.3.4 Size of margins

HELP example: see 3.7.3

Within SAS, the margin options define the printable area of the page for graphics and text. For R, these control how tight plots are to the printable area.

SAS

```
options bottommargin=3in topmargin=4cm leftmargin=1 rightmargin=1;
```

Note: The default units are inches; a trailing `cm` indicates centimeters, while a trailing `in` makes inches the explicit metric.

R

```
par(mar=c(bot, left, top, right),    # inner margin
    oma=c(bot, left, top, right))    # outer margin
```

Note: The vector given to `mar` specifies the number of lines of margin around a plot: the default is `c(5, 4, 4, 2) + 0.1`. The `oma` option specifies additional lines outside the entire plotting area (the default is `c(0,0,0,0)`). Other options to control margin spacing include `omd` and `omi`.

5.3.5 Graphical settings

HELP example: see 3.7.3

SAS

```
goptions reset=all;
```

Note: Many graphical settings are specified using the `goptions` statement. The above usage will revert all values to the SAS defaults.

R

```
# change values, while saving old
oldvalues <- par(...)

# restore old values for graphical settings
par(oldvalues)
```

5.3.6 Multiple plots per page

HELP example: see 3.7.3

In SAS, putting multiple arbitrary plots onto a page is possible but is non-trivial and is beyond the scope of the book. Examples can be found in the on-line help for `proc greplay`: Contents; SAS Products; SAS Procedures; Proc Greplay. Scatterplot matrices (5.1.17) can be generated using `proc sgscatter` and conditioning plots (5.1.11) can be made using `proc sgpanel`.

R

```
par(mfrow=c(a, b))
```

or

```
par(mfcol=c(a, b))
```

Note: The `mfrow` option specifies that plots will be drawn in an a × b array by row (by column for `mfcol`).

5.3.7 Axis range and style

SAS

HELP example: see 3.7.1, 5.6.1

```
axis1 order = (x1, x2 to x3 by x4, x5);
axis2 order = ("value1" "value2" ... "valuen");
```

Note: Axis statements are associated with vertical or horizontal axes using `vaxis` or `haxis` options in various procedures. For an example, see Figure 5.1 in section 5.6.1. Multiple options to the `axis` statement can be listed, as in Figure 3.1. The `axis` statement does not apply to most `ODS graphics` (A.7.3) output.

R

```
plot(x, y, xlim=c(minx, maxx), ylim=c(miny, maxy), xaxs="i", yaxs="i")
```

Note: The `xaxs` and `yaxs` options control whether tick marks extend beyond the limits of the plotted observations (default) or are constrained to be internal (`"i"`). More control is available through the `axis()` and `mtext()` functions.

5.3.8 Axis labels, values, and tick marks

HELP example: see 1.13.5

SAS

```
axis1 label=("Text for axis label" angle=90 color=red
   font=swiss height=2 justify=right rotate=180);
axis1 value=("label1" "label2")
axis1 major=(color=blue height=1.5cm width=2);
axis1 minor=none;
```

Note: Axis statements are associated with vertical or horizontal axes using `vaxis` or `haxis` options in various procedures. For example, in `proc gplot`, one might use a `plot y*x / vaxis=axis1 haxis=axis2` statement. Multiple options to the `axis` statement can be listed, as in Figure 3.1. The `axis` statement does not apply to most `ODS graphics` (A.7.3) output.

In the `label` option above we show the text options available for graphics which apply to both `legend` and `axis` statements, and to `title` statements when graphics are produced. The `angle` option specifies the angle of the line along which the text is printed; the default depends on which `axis` is described. The `color` and `font` options are discussed in sections 5.3.11 and 5.2.11, respectfully. The `height` option specifies the text size; it is measured in graphic cells, but can be specified with the number of units, for example `height=1cm`. The `justify` option can take values of `left`, `center`, or `right`. The `rotate` option rotates each character in place. The `value` option describes the text which labels the tick marks, and takes the same parameters described for the `label` option.

The `major` and `minor` options take the same parameters; `none` will omit either labeled (`major`) or unlabeled (`minor`) tick marks. The `width` option specifies the thickness of the tick in multiples of the default.

R

```
plot(x, y, lab=c(x, y, len),  # number of tick marks
   las=lasval,     # orientation of tick marks
   tck=tckval,     # length of tick marks
   tcl=tclval,     # length of tick marks
   xaxp=c(x1, x2, n),    # coordinates of the extreme tick marks
   yaxp=c(x1, x2, n),    # coordinates of the extreme tick marks
   xlab="X axis label", ylab="Y axis label")
```

Note: Options for `las` include `0` for always parallel, `1` for always horizontal, `2` for perpendicular, and `3` for vertical.

5.3.9 Line styles

HELP example: see 3.7.3

SAS

```
symbol1 interpol=itype line=ltyval;
```

Note: The `interpol` option to the `symbol` statement, which can be shortened to simply `i`, specifies what kind of line should be plotted through the data. Options include smoothers, step functions, linear regressions, and more. The `line` option (which can be shortened to `l`) specifies a solid line (by default, `ltyval=1`) or various dashed or dotted lines (`ltyval 2 ... 33`). A list of line types with associated code can be found in the on-line documentation: Contents; SAS Products; SAS/GRAPH; Procedures and Statements; Symbol. The line types do not have a separate entry, but appear near the end of the long description of the `symbol` statement.

R

```
plot(...)
lines(x, y, lty=ltyval)
```

Note: Supported line type values include 0=blank, 1=solid (default), 2=dashed, 3=dotted, 4=dotdash, 5=longdash, and 6=twodash.

5.3.10 Line widths

SAS *HELP example:* see 1.13.5

```
symbol interpol=interpol_type width=lwdval;
```

Note: When a line through the data is requested using the `interpol` option, the thickness of the line, in multiples of the default thickness, can be specified by the `width` option, for which `w` is a synonym. The default thickness depends on display hardware.

R

```
plot(...)
lines(x, y, lwd=lwdval)
```

Note: The default for `lwd` is 1; the value of `lwdval` must be positive.

5.3.11 Colors

SAS *HELP example:* see 2.6.4

```
symbol1 c=colval cl=colval cv=colval;
axis1 label=(color=colval);
```

Note: Colors in SAS can be specified in a variety of ways. Some typical examples of applying colors are shown, but many features of plots can be colored. If precise control is required, `colval` can be specified using a variety of schemes as described in the on-line documentation: Contents; SAS Products; SAS/GRAPH; Concepts; Colors. For more casual choice of colors, color names such as `blue`, `black`, `red`, `purple`, `strongblue`, or `lightred` can be used.

R

```
plot(...)
lines(x, y, col=colval)
```

Note: For more information on setting colors, see the `Color Specification` section within `help(par)`. The `colors()` function lists available colors, while `colors.plot()` function within the `epitools` package displays a matrix of colors, and `colors.matrix()` returns a matrix of color names. The `display.brewer.all()` function within the `RColorBrewer` package is particularly useful for selecting a set of complementary colors for a palette.

5.3.12 Log scale

SAS

```
axis1 logbase=base logstyle=expand;
```

Note: The `logbase` option scales the axis according to the log of the specified base; valid base values include `e`, `pi`, or a number. The `logstyle` option produces plots with tick marks labeled with the value of the base (`logstyle=power`) or the base raised to that value (`logstyle=expand`).

R

```
plot(x, y, log=logval)
```

Note: A natural log scale can be specified using the `log` option to `plot()`, where `log="x"` denotes only the x axis, `"y"` only the y axis, and `"xy"` for both.

5.3.13 Omit axes

SAS *HELP example:* see 6.4

```
axis1 style=0 major=none minor=none label=("") value=none;
```

Note: To remove an axis entirely in SAS, it is necessary to request each element of the axis not be drawn, as shown here.

R

```
plot(x, y, xaxt="n", yaxt="n")
```

5.4 Saving graphs

It is straightforward to export graphics in a variety of formats. In SAS, this can be done using the `ODS` system or via the `goptions` statement. The former will integrate procedure output and graphics. The latter is more cumbersome and cannot be used with `ODS` graphics or the `sgplot`, `sgpanel`, or `sgscatter` procedures. However, it supports more formats, and will work with `gplot`, `gchart`, and other SAS/GRAPH procedures.

5.4.1 PDF

SAS *HELP example:* see 6.4.2

```
ods pdf file="filename.pdf";
proc gplot data=ds;
   ...
ods pdf close;
```

or

```
filename filehandle "filename.pdf";
goptions gsfname=filehandle device=pdf gsfmode=replace;

proc gplot data=ds;
   ...
run;
```

Note: In both versions above, the `filename` can include a directory location as well as a name. The `device` option specifies formatting of the graphic; the many valid options can be viewed using `proc gdevice` and key options are presented in this section. The `gsfmode=replace` option allows SAS to create and/or overwrite the graphic. The `filehandle` basename may not be more than 8 characters long.

The `ods pdf` statement will place graphics and text output from procedures into the pdf file generated.

R

```
pdf("file.pdf")
plot(...)
dev.off()
```

Note: The `dev.off()` function is used to close a graphics device.

5.4.2 Postscript

SAS

```
ods ps file="filename.ps";
proc gplot data=ds;
   ...
run;
ods ps close;
```

or

```
filename filehandle "filename.ps";
goptions gsfname=filehandle device=ps gsfmode=replace;

proc gplot data=ds;
   ...
run;
```

Note: In both versions above, the `filename` can include a directory location as well as a name. The `device` option specifies formatting of the graphic; the many valid options can be viewed using `proc gdevice` and key options are presented in this section. The `gsfmode=replace` option allows SAS to create and/or overwrite the graphic. The `filehandle` basename may not be more than 8 characters long.

The `ods ps` statement will place graphics and text output from procedures into the pdf file generated.

R

```
postscript("file.ps")
plot(...)
dev.off()
```

Note: The `dev.off()` function is used to close a graphics device.

5.4.3 RTF

The Rich Text Format (RTF) is a file format developed for cross-platform document sharing. Most word processors are able to read and write RTF documents. The following will create a file in this format containing the graphic; any text generated by procedures will also appear in the RTF file if they are executed between the `ods rtf` and `ods rtf close` statements. We are not aware of a similar capability in R.

SAS

```
ods rtf file="filename.rtf";
proc gplot data=ds;
   ...
run;
ods rtf close;
```

Note: The `filename` can include a directory location as well as a name.

5.4.4 JPEG

SAS

```
filename filehandle "filename.jpg";
goptions gsfname=filehandle device=jpeg gsfmode=replace;

proc gplot data=ds;
   ...
run;
```

Note: The `filename` can include a directory location as well as a name. The `device` option specifies formatting of the graphic; valid options can be viewed using `proc gdevice`. The `gsfmode=replace` option allows SAS to create and/or overwrite the graphic. The `filehandle` basename may not be more than 8 characters long.

R

```
jpeg("filename.jpg")
plot(...)
dev.off()
```

Note: The `dev.off()` function is used to close a graphics device.

5.4.5 WMF

SAS

```
filename filehandle "filename.wmf";
goptions gsfname=filehandle device=wmf gsfmode=replace;

proc gplot data=ds;
   ...
run;
```

Note: The `filename` can include a directory location as well as a name. The `device` option specifies formatting of the graphic; valid options can be viewed using `proc gdevice`. The `gsfmode=replace` option allows SAS to create and/or overwrite the graphic. The `filehandle` basename may not be more than 8 characters long.

R

```
win.metafile("file.wmf")
plot(...)
dev.off()
```

Note: The function `win.metafile()` is only supported under Windows. Functions which generate multiple plots are not supported. The `dev.off()` function is used to close a graphics device.

5.4.6 BMP

SAS

```
filename filehandle "filename.bmp";
goptions gsfname=filehandle device=bmp gsfmode=replace;

proc gplot data=ds;
   ...
run;
```

Note: The `filename` can include a directory location as well as a name. The `device` option specifies formatting of the graphic; valid options can be viewed using `proc gdevice`. The `gsfmode=replace` option allows SAS to create and/or overwrite the graphic. The `filehandle` basename may not be more than 8 characters long.

R

```
bmp("filename.bmp")
plot(...)
dev.off()
```

Note: The `dev.off()` function is used to close a graphics device.

5.4.7 TIFF

SAS

```
filename filehandle "filename.tif";
goptions gsfname=filehandle device=tiffp300 gsfmode=replace;

proc gplot data=ds;
   ...
run;
```

Note: The `filename` can include a directory location as well as a name. The `device` option specifies formatting of the graphic; valid options can be viewed using `proc gdevice`. The `gsfmode=replace` option allows SAS to create and/or overwrite the graphic. The `filehandle` basename may not be more than 8 characters long. Many types of TIFF can be generated; the above `device` specifies a color plot with 300 dpi.

R

```
tiff("filename.tiff")
plot(...)
dev.off()
```

Note: The `dev.off()` function is used to close a graphics device.

5.4.8 PNG

SAS

```
filename filehandle "filename.png";
goptions gsfname=filehandle device=png gsfmode=replace;

proc gplot data=ds;
   ...
run;
```

Note: The `filename` can include a directory location as well as a name. The `device` option specifies formatting of the graphic; valid options can be viewed using `proc gdevice`. The `gsfmode=replace` option allows SAS to create and/or overwrite the graphic. The `filehandle` basename may not be more than 8 characters long.

The `ODS` graphics system works by creating a PNG file which is stored in the current directory or the directory and then creating output in the desired format. So using the `ods output` statement for of the formatted output options in this section will also result in a PNG file.

R

```
png("filename.png")
plot(...)
dev.off()
```

Note: The `dev.off()` function is used to close a graphics device.

5.4.9 Closing a graphic device

HELP example: see 5.6.3

There is no analog in SAS for this concept. In R, the following code closes a graphics window. This is particularly useful when a graphics file is being created.

R

```
dev.off()
```

5.5 Further resources

The books by Tufte [87, 88, 89, 90] provide an excellent framework for graphical displays, some of which build on the work of Tukey [91]. Comprehensive and accessible books on R graphics include [56] and [74].

5.6 HELP examples

To help illustrate the tools presented in this chapter, we apply many of the entries to the HELP data. SAS and R code can be downloaded from `http://www.math.smith.edu/sasr/examples`. We begin by reading in the data.

```
proc import
   datafile='c:/book/help.csv'
   out=ds
   dbms=dlm;
   delimiter=',';
   getnames=yes;
run;
```

```
> options(digits=3)
> ds <- read.csv("help.csv")
> attach(ds)
```

5.6.1 Scatterplot with multiple axes

The following example creates a single Figure that displays the relationship between CESD and the variables `indtot` (Index of Drug Abuse Consequences, InDUC) and `mcs` (Mental Component Score), for a subset of female alcohol-involved subjects. We specify two different y-axes (5.1.2) for the Figure.

```
axis1 minor=none;
axis2 minor=none order=(5 to 60 by 13.625);
axis3 minor=none order=(20, 40, 60);
symbol1 i=sm65s v=circle color=black l=1 w=5;
symbol2 i=sm65s v=triangle color=black l=2 w=5;
proc gplot data=ds;
   where female eq 1 and substance eq 'alcohol';
   plot indtot*cesd / vaxis=axis1 haxis=axis3;
   plot2 mcs*cesd / vaxis = axis2;
run; quit;
```

In the SAS code above, the `symbol` and axis `statements` are used to control the output and to add lines through the data. Note that three axes are specified and are associated with the various axes in the plot in the `vaxis` and `haxis` options to the `plot` and `plot2` statements. The `axis` statements can be omitted for a simpler graphic.

In R, a considerable amount of housekeeping is needed. The second y variable must be rescaled to the range of the original, and the axis labels and tick marks added on the right. To accomplish this, we write a function `plottwoy()` which first makes the plot of the first (left axis) y against x, adds a lowess curve through that data, then calls a second function, `addsecondy()`.

```
> plottwoy <- function(x, y1, y2, xname="X", y1name="Y1", y2name="Y2")
+ {
+    plot(x, y1, ylab=y1name, xlab=xname)
+    lines(lowess(x, y1), lwd=3)
+    addsecondy(x, y2, y1, yname=y2name)
+ }
```

The function `addsecondy()` does the work of rescaling the range of the second variable to that of the first, adds the right axis, and plots a lowess curve through the data for the rescaled `y2` variable.

```
> addsecondy <- function(x, y, origy, yname="Y2") {
+    prevlimits <- range(origy)
+    axislimits <- range(y)
+    axis(side=4, at=prevlimits[1] + diff(prevlimits)*c(0:5)/5,
+       labels=round(axislimits[1] + diff(axislimits)*c(0:5)/5, 1))
+    mtext(yname, side=4)
+    newy <- (y-axislimits[1])/(diff(axislimits)/diff(prevlimits)) +
+       prevlimits[1]
+    points(x, newy, pch=2)
+    lines(lowess(x, newy), lty=2, lwd=3)
+ }
```

Finally, the newly defined functions can be run and Figure 5.1 generated.

```
> plottwoy(cesd[female==1&substance=="alcohol"],
+    indtot[female==1&substance=="alcohol"],
+    mcs[female==1&substance=="alcohol"], xname="cesd",
+    y1name="indtot", y2name="mcs")
```

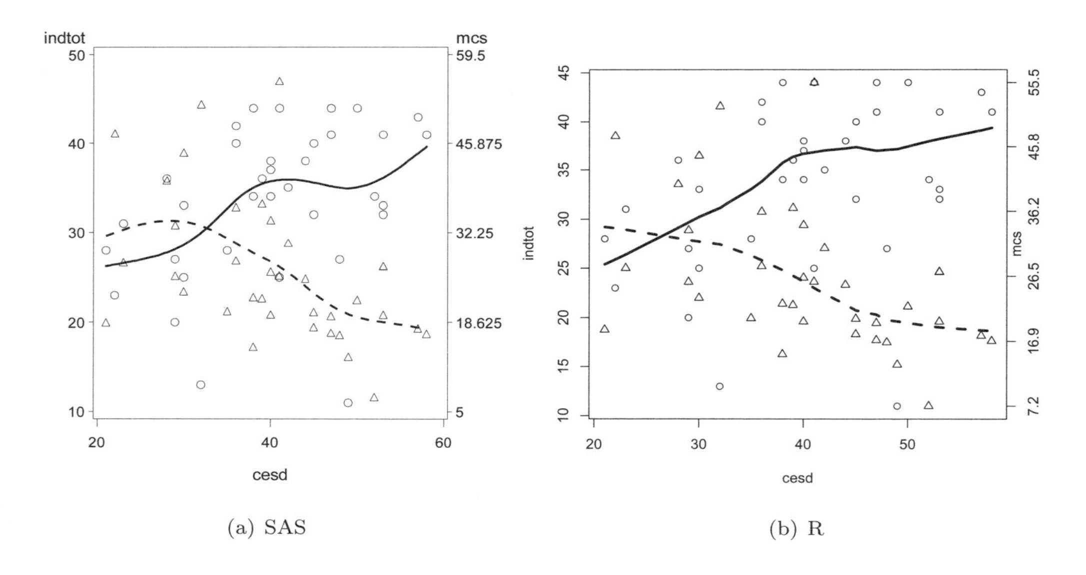

(a) SAS

(b) R

Figure 5.1: Plot of InDUC and MCS vs. CESD for female alcohol-involved subjects

Note that the two graphics appear to be different due to different right y-axes. In SAS it is difficult to select axis ranges exactly conforming to the range of the data, while our R function uses more of the space for data display.

5.6.2 Conditioning plot

Figure 5.2 displays a conditioning plot (5.1.11) with the association between MCS and CESD stratified by substance and report of suicidal thoughts (`g1b`).

Note that SAS version 9.2 is required; the plot is hard to replicate with earlier versions of SAS.

```
proc sgpanel data=ds;
   panelby g1b substance / layout=lattice;
   pbspline x=cesd y=mcs;
run; quit;
```

For R, ensure that the necessary packages are installed (B.6.1).

```
> library(lattice)
```

Then set up and generate the plot.

```
> suicidal.thoughts <- as.factor(g1b)
> coplot(mcs ~ cesd | suicidal.thoughts*substance,
+    panel=panel.smooth)
```

There is a similar association between CESD and MCS for each of the substance groups. Subjects with suicidal thoughts tended to have higher CESD scores, and the association between CESD and MCS was somewhat less pronounced than for those without suicidal thoughts.

5.6.3 Kaplan–Meier plot

The main outcome of the HELP study was time to linkage to primary care, as a function of randomization group. This can be displayed using a Kaplan–Meier plot (see 5.1.19). For SAS detailed information regarding the Kaplan–Meier estimator at each time point can be found by omitting the `ods select` statement; for R by using `summary(survobj)`. Figure 5.3 displays the estimates, with + signs indicating censored observations.

```
ods graphics on;
ods select censoredsummary survivalplot;
proc lifetest data=ds plots=s(test);
   time dayslink*linkstatus(0);
   strata treat;
run;
ods graphics off;
```

```
The LIFETEST Procedure

      Summary of the Number of Censored and Uncensored Values

                                                              Percent
Stratum            treat         Total   Failed     Censored    Censored

      1                0           209       35          174       83.25
      2                1           222      128           94       42.34
---------------------------------------------------------------------
  Total                            431      163          268       62.18

NOTE: 22 observations with invalid time, censoring, or strata values were
deleted.
```

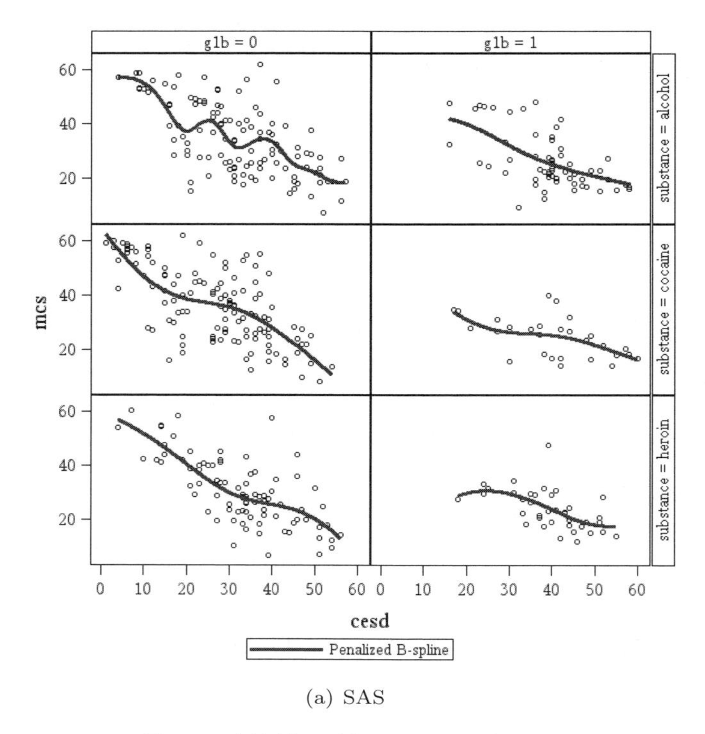

(a) SAS

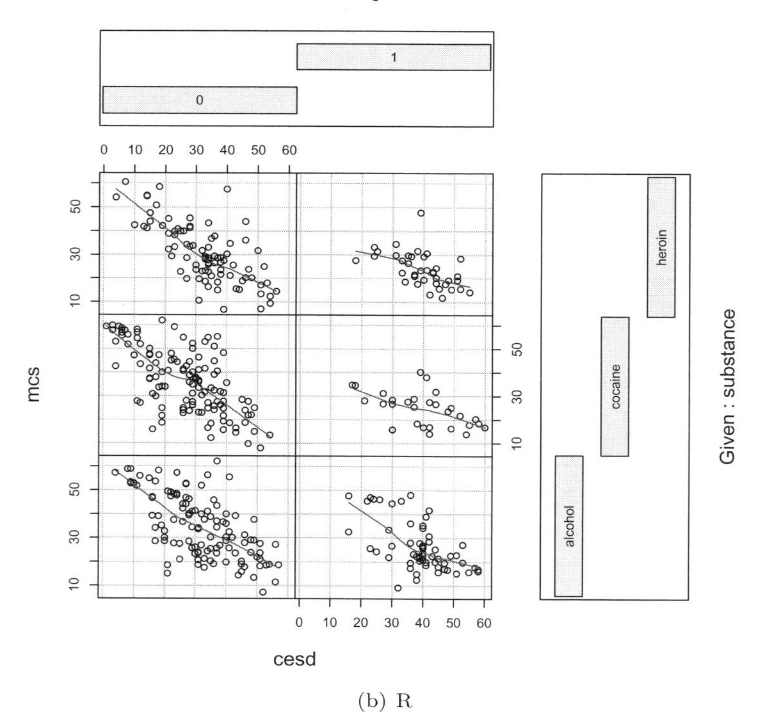

(b) R

Figure 5.2: Association of MCS and CESD, stratified by substance and report of suicidal thoughts

```
> library(survival)
> survobj <- survfit(Surv(dayslink, linkstatus) ~ treat)
> print(survobj)

Call: survfit(formula = Surv(dayslink, linkstatus) ~ treat)

   22 observations deleted due to missingness
          n events median 0.95LCL 0.95UCL
treat=0 209     35    Inf     Inf     Inf
treat=1 222    128    120      79     272
```

```
> plot(survobj, lty=1:2, lwd=2, col=c(4,2))
> title("Product-Limit Survival Estimates")
> legend(250, .75, legend=c("Control", "Treatment"), lty=c(1,2), lwd=2,
+     col=c(4,2), cex=1.4)
```

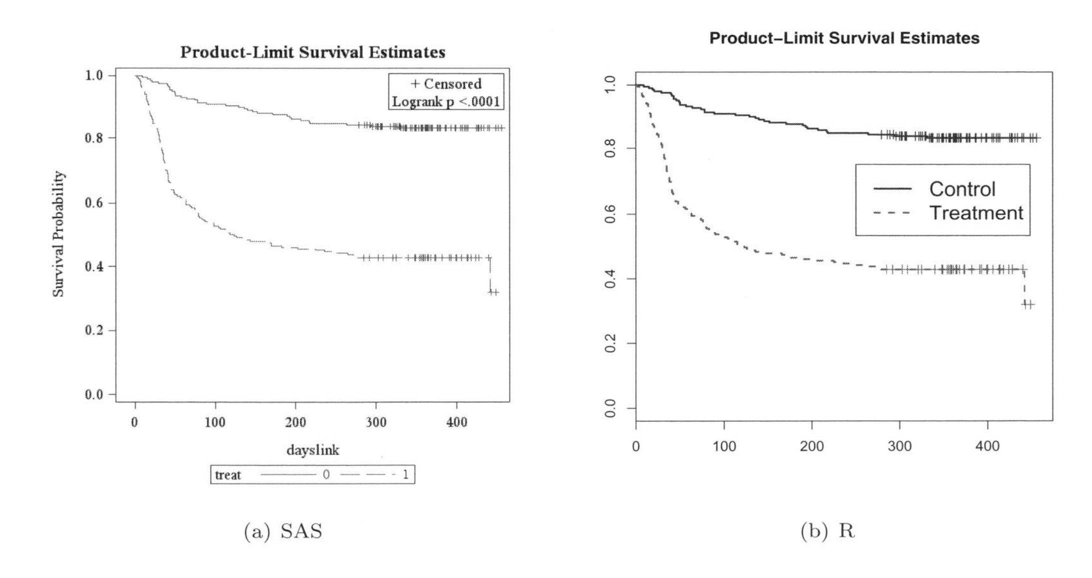

(a) SAS

(b) R

Figure 5.3: Kaplan–Meier estimate of time to linkage to primary care by randomization group

As reported previously [35, 71], there is a highly statistically significant effect of treatment, with approximately 55% of clinic subjects linking to primary care, as opposed to only 15% of control subjects.

5.6.4 ROC curve

Receiver operating characteristic (ROC) curves are used for diagnostic agreement (2.2.2 and 5.1.18) as well as assessing goodness of fit for logistic regression (4.1.1). In SAS, they can be created using `proc logistic`. In R, we use the `ROCR` library. Figure 5.4 displays the receiver operating characteristic curve predicting suicidal thoughts using the CESD measure of depressive symptoms.

```
ods graphics on;
ods select roccurve;
proc logistic data=ds descending plots(only)=roc;
   model g1b = cesd;
run;
ods graphics off;
```

The `descending` option changes the behavior of `proc logistic` to model the probability that the outcome is 1; the default models the probability that the outcome is 0.

Using R, we first load the `ROCR` library, create a prediction object, and retrieve the area under the curve (AUC) to use in Figure 5.4.

```
> library(ROCR)
> pred <- prediction(cesd, g1b)
> auc <- slot(performance(pred, "auc"), "y.values")[[1]]
```

We can then plot the ROC curve, adding display of cutoffs for particular CESD values ranging from 20 to 50. These values are offset from the ROC curve using the `text.adj` option.

If the continuous variable (in this case `cesd`) is replaced by the predicted probability from a logistic regression model, multiple predictors can be included.

```
> plot(performance(pred, "tpr", "fpr"),
+    print.cutoffs.at=seq(from=20, to=50, by=5),
+    text.adj=c(1, -.5), lwd=2)
> lines(c(0, 1), c(0, 1))
> text(.6, .2, paste("AUC=", round(auc,3), sep=""), cex=1.4)
> title("ROC Curve for Model")
```

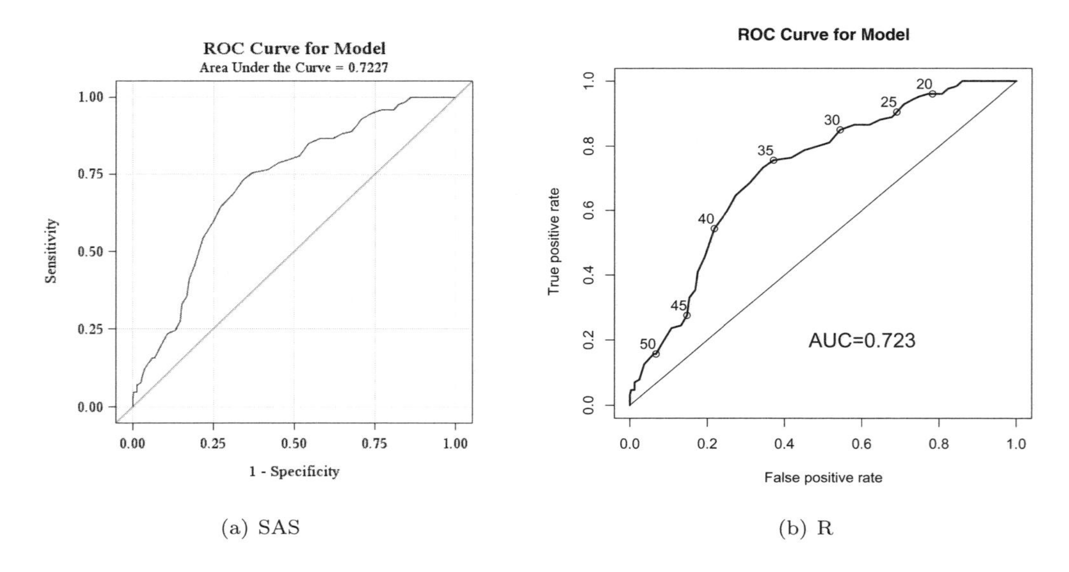

(a) SAS

(b) R

Figure 5.4: Receiver operating characteristic curve for the logistical regression model predicting suicidal thoughts using the CESD as a measure of depressive symptoms (sensitivity = true positive rate; 1-specificity = false positive rate)

5.6.5 Pairs plot

We can qualitatively assess the associations between some of the continuous measures of mental health, physical health, and alcohol consumption using a pairsplot or scatterplot matrix (5.1.17). To make the results clearer, we display only the female subjects.

For SAS, the new `sgscatter` procedure provides a simple way to produce this. The results of the following code are included in Figure 5.5.

```
proc sgscatter data=ds;
   where female eq 1;
   matrix cesd mcs pcs i1 / diagonal=(histogram kernel);
run; quit;
```

If fits in the pairwise scatterplots are required, the following code will produce a similar matrix, with loess curves in each cell and less helpful graphs in the diagonals (results not shown).

```
proc sgscatter data=ds;
   where female eq 1;
   compare x = (cesd mcs pcs i1)
           y = (cesd mcs pcs i1) / loess;
run; quit;
```

For complete control of the figure, the `sgscatter` procedure will not suffice and more complex coding is necessary; we would begin with SAS macros written by Michael Friendly and available from his web site at York University.

For R, a simple version with only the scatterplots could be generated easily with the `pairs()` function (results not shown):

```
> pairs(c(ds[72:74], ds[67]))
```

or

```
> pairs(ds[c("pcs", "mcs", "cesd", "i1")])
```

Here instead we demonstrate building a figure using several functions. We begin with a function `panel.hist()` to display the diagonal entries (in this case, by displaying a histogram).

```
> panel.hist <- function(x, ...)
+ {
+     usr <- par("usr"); on.exit(par(usr))
+     par(usr = c(usr[1:2], 0, 1.5) )
+     h <- hist(x, plot=FALSE)
+     breaks <- h$breaks; nB <- length(breaks)
+     y <- h$counts; y <- y/max(y)
+     rect(breaks[-nB], 0, breaks[-1], y, col="cyan", ...)
+ }
```

Another function is created to create a scatterplot along with a fitted line.

```
> panel.lm <- function(x, y, col=par("col"), bg=NA, pch=par("pch"),
+     cex=1, col.lm="red", ...)
+ {
+     points(x, y, pch=pch, col=col, bg=bg, cex=cex)
+     ok <- is.finite(x) & is.finite(y)
+     if (any(ok))
+         abline(lm(y[ok] ~ x[ok]))
+ }
```

These functions are called (along with the built-in `panel.smooth()` function) to display the results. Figure 5.5 displays the pairsplot of CESD, MCS, PCS, and I1, with histograms along the diagonals. For R, smoothing splines are fit on the lower triangle, linear fits on the upper triangle, using code fragments derived from `example(pairs)`.

```
> pairs(~ cesd + mcs + pcs + i1, subset=(female==1),
+    lower.panel=panel.smooth, diag.panel=panel.hist,
+    upper.panel=panel.lm)
```

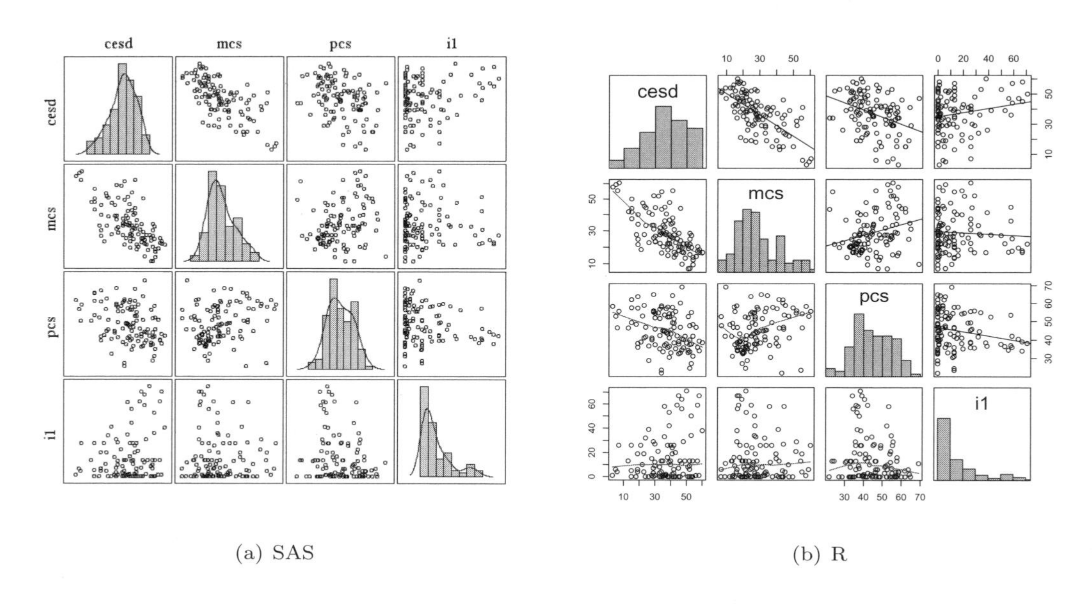

(a) SAS (b) R

Figure 5.5: Pairsplot of variables from the HELP dataset

There is an indication that `CESD`, `MCS`, and `PCS` are interrelated, while `I1` appears to have modest associations with the other variables.

5.6.6 Visualize correlation matrix

One visual analysis which might be helpful to display would be the pairwise correlations. We approximate this in SAS by plotting a confidence ellipse for the observed data. This approach allows an assessment of whether the linear correlation is an appropriate statistic to consider.

In the code below, we demonstrate some options for the `sgscatter procedure`. The `ellipse` option draws confidence ellipses at the requested α-level; here chosen arbitrarily to mimic R. The `start` option also mimics R by making the diagonal begin in the lower left; the top left is the default. The `markerattrs` option controls aspects of the appearance of plots generated with the `sgscatter`, `sgpanel`, and `sgplot` procedures.

```
proc sgscatter data=ds;
   matrix mcs pcs pss_fr drugrisk cesd indtot i1 sexrisk /
      ellipse=(alpha=.25) start=bottomleft
      markerattrs=(symbol=circlefilled size=2);
run; quit;
```

In R, we utilize the approach used by Sarkar to recreate Figure 13.5 of the *Lattice: Multivariate data visualization with R* book [74]. Other examples in that reference help to motivate the power of the `lattice` package far beyond what is provided by `demo(lattice)`.

```
> cormat <- cor(cbind(mcs, pcs, pss_fr, drugrisk, cesd, indtot, i1,
+    sexrisk), use="pairwise.complete.obs")
> oldopt <- options(digits=2)
> cormat

            mcs    pcs pss_fr drugrisk   cesd indtot     i1 sexrisk
mcs       1.000  0.110  0.138  -0.2058 -0.682  -0.38 -0.087 -0.1061
pcs       0.110  1.000  0.077  -0.1411 -0.293  -0.13 -0.196  0.0239
pss_fr    0.138  0.077  1.000  -0.0390 -0.184  -0.20 -0.070 -0.1128
drugrisk -0.206 -0.141 -0.039   1.0000  0.179   0.18 -0.100 -0.0055
cesd     -0.682 -0.293 -0.184   0.1789  1.000   0.34  0.176  0.0157
indtot   -0.381 -0.135 -0.198   0.1807  0.336   1.00  0.202  0.1132
i1       -0.087 -0.196 -0.070  -0.0999  0.176   0.20  1.000  0.0881
sexrisk  -0.106  0.024 -0.113  -0.0055  0.016   0.11  0.088  1.0000

> options(oldopt)
```

```
> drugrisk[is.na(drugrisk)] <- 0
> panel.corrgram <- function(x, y, z, at, level=0.9, label=FALSE, ...)
+ {
+    require("ellipse", quietly=TRUE)
+    zcol <- level.colors(z, at=at, col.regions=gray.colors)
+    for (i in seq(along=z)) {
+       ell <- ellipse(z[i], level=level, npoints=50,
+          scale=c(.2, .2), centre=c(x[i], y[i]))
+       panel.polygon(ell, col=zcol[i], border=zcol[i], ...)
+    }
+    if (label)
+       panel.text(x=x, y=y, lab=100*round(z, 2), cex=0.8,
+          col=ifelse(z < 0, "white", "black"))
+ }
```

```
> library(ellipse)
> library(lattice)
> print(levelplot(cormat, at=do.breaks(c(-1.01, 1.01), 20),
+    xlab=NULL, ylab=NULL, colorkey=list(space = "top",
+    col=gray.colors), scales=list(x=list(rot = 90)),
+    panel=panel.corrgram,
+    label=TRUE))
```

The SAS plot suggests that some of these linear correlations might not be useful measures of association, while the R plot allows a consistent frame of reference for the many correlations.

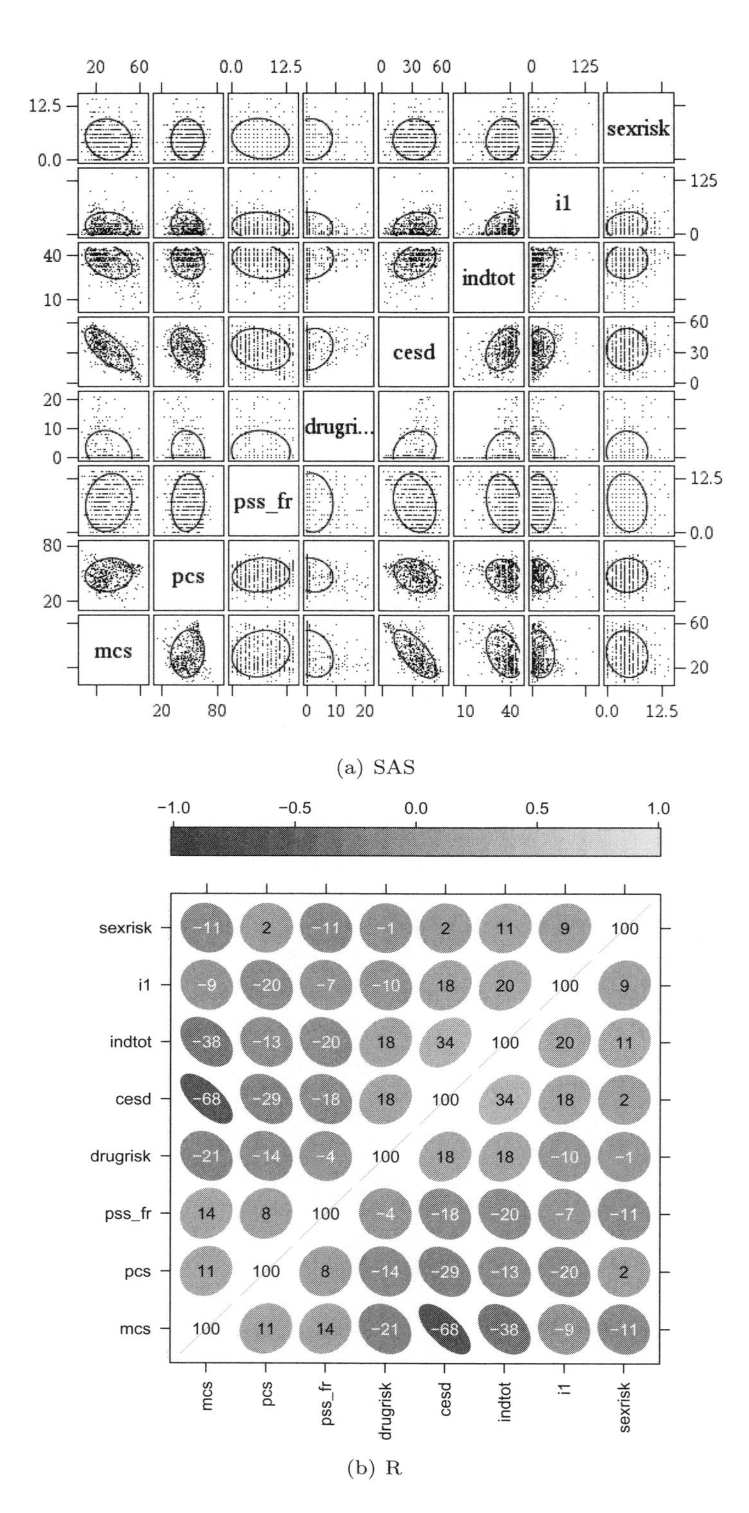

(a) SAS

(b) R

Figure 5.6: Visual display of correlations and associations

Chapter 6

Other topics and extended examples

In this chapter, we address several additional topics and extended examples through parallel implementations that show off the statistical computing strengths and potential of these two packages. In addition, we briefly describe implementations that facilitate power calculations, multivariate procedures, handling of missing data, fitting Bayesian models, and accounting for complex survey design.

6.1 Power and sample size calculations

Many simple settings lend themselves to analytic power calculations, where closed form solutions are available. Other situations may require an empirical calculation, where repeated simulation is undertaken.

6.1.1 Analytic power calculation

It is straightforward to find power or sample size (given a desired power) for two sample comparisons of either continuous or categorical outcomes. We show simple examples for comparing means and proportions in two groups and supply additional information on analytic power calculation available for more complex methods.

SAS

```
/* find sample size for two-sample t-test */
proc power;
   twosamplemeans groupmeans=(0 0.5) stddev=1 power=0.9 ntotal=.;
run;
```

```
/* find power for two-sample t-test */
proc power;
   twosamplemeans groupmeans=(0 0.5) stddev=1 power=. ntotal=200;
run;
```

The latter call generates the following output:

```
The POWER Procedure
Two-sample t Test for Mean Difference
     Fixed Scenario Elements
Distribution                 Normal
Method                        Exact
Group 1 Mean                      0
Group 2 Mean                    0.5
Standard Deviation                1
Total Sample Size               200
Number of Sides                   2
Null Difference                   0
Alpha                          0.05
Group 1 Weight                    1
Group 2 Weight                    1

Computed Power
Power 0.940
```

```
/* find sample size for two-sample test of proportions */
proc power;
   twosamplefreq test=pchi ntotal=. groupproportions=(.1 .2) power=0.9;
run;
```

```
/* find power for two-sample test of proportions */
proc power;
   twosamplefreq test=pchi ntotal=200 groupproportions=(.1 .2) power=.;
run;
```

Note: The power procedure also allows power calculations for the Wilcoxon rank-sum test, the log-rank and related tests for censored data, paired tests of means and proportions, correlations, and for ANOVA and linear and logistic regression. The syntax is similar with the desired output of power, total sample size, effect size, alpha level, or variance listed with a missing value (a period after the equals sign).

R

```
# find sample size for two-sample t-test
power.t.test(delta=0.5, power=0.9)
```

```
# find power for two-sample t-test
power.t.test(delta=0.5, n=100)
```

The latter call generates the following output:

```
     Two-sample t test power calculation
              n = 100
          delta = 0.5
             sd = 1
      sig.level = 0.05
          power = 0.9404272
    alternative = two.sided
 NOTE: n is number in *each* group
```

```
# find sample size for two-sample test of proportions
power.prop.test(p1=.1, p2=.2, power=.9)
```

```
# find power for two-sample test of proportions
power.prop.test(p1=.1, p2=.2, n=100)
```

Note: The `power.t.test()` function requires exactly four of the five arguments (sample size in each group, power, difference between groups, standard deviation, and significance level) to be specified. Default values exist for `sd=1` and `sig.level=0.05`. Other power calculation functions can be found in the `pwr` package.

6.1.2 Simulation-based power calculations

In some settings, analytic power calculations may not be readily available. A straightforward alternative is to estimate power empirically, simulating data from the proposed design under given assumptions regarding the alternative.

We consider a study of children clustered within families. Each family has 3 children; in some families all 3 children have an exposure of interest, while in others just 1 child is exposed. In the simulation, we assume that the outcome is multivariate normal with higher mean for those with the exposure, and 0 for those without. A compound symmetry correlation is assumed, with equal variances at all times. We assess the power to detect an exposure effect where the intended analysis uses a random intercept model (4.2.2) to account for the clustering within families.

With this simple covariance structure it is trivial to generate correlated errors directly, as in the SAS code below; an alternative which could be used with more complex structures in SAS would be `proc simnorm` (1.10.6).

SAS

```
data simpower1;
   effect = 0.35;  /* effect size */
   corr = 0.4;     /* desired correlation */
   covar = (corr)/(1 - corr);  /* implied covariance given variance = 1*/
   numsim = 1000;  /* number of datasets to simulate */
   numfams = 100;  /* number of families in each dataset */
   numkids = 3;    /* each family */
   do simnum = 1 to numsim;    /* make a new dataset for each simnum */
      do famid = 1 to numfams; /* make numfams families in each dataset */
         inducecorr = normal(42)* sqrt(covar);
                     /* this is the mechanism to achieve the desired
                                correlation between kids within family */
         do kidnum = 1 to numkids;    /* generate each kid */
            exposed = ((kidnum eq 1) or (famid le numfams/2)) ;
                                /* assign kid to be exposed */
            x = (exposed * effect) +
                (inducecorr + normal(0))/sqrt(1 + covar);
            output;
         end;
      end;
   end;
run;
```

In the code above, the integer provided as an argument in the initial use of the `normal` function sets the seed used for all calls to the pseudo-random number generator, so that the results can be exactly replicated, if necessary (see section 1.10.9.) Next, we run the desired model on each of the simulated datasets, using the `by` statement (A.6.2) and saving the estimated fixed effects parameters using the `ODS` system (A.7).

```
ods select none;
ods output solutionf=simres;
proc mixed data=simpower1 order=data;
by simnum;
   class exposed famid;
   model x = exposed / solution;
   random int / subject=famid;
run;
ods select all;
```

Finally, we process the resulting output dataset to generate an indicator of rejecting the null hypothesis of no exposure effect.

```
data powerout;
set simres;
   where exposed eq 1;
   reject=(probt lt 0.05);
run;
```

Note: The proportion of rejections shown in the results of `proc freq` is the empirical estimate of power.

```
proc freq data=powerout;
   tables reject / binomial (level='1');
run;
```

```
The FREQ Procedure
                                          Cumulative    Cumulative
reject    Frequency      Percent      Frequency       Percent
-----------------------------------------------------------------
     0         153        15.30            153         15.30
     1         847        84.70           1000        100.00
```

The `binomial` option to `proc freq` provides asymptotic and exact CI for this estimated power:

```
Proportion                  0.8470
ASE                         0.0114
95% Lower Conf Limit        0.8247
95% Upper Conf Limit        0.8693
```

```
Exact Conf Limits
95% Lower Conf Limit        0.8232
95% Upper Conf Limit        0.8688
```

In R, we specify the correlation matrix directly, and simulate the multivariate normal.

R

```
library(MASS)
library(nlme)
# initialize parameters and building blocks
effect <- 0.35     # effect size
corr <- 0.4        # intrafamilial correlation
numsim <- 1000
n1fam <- 50        # families with 3 exposed
n2fam <- 50        # families with 1 exposed and 2 unexposed
vmat <- matrix(c  # 3x3 compound symmetry correlation
   (1,    corr, corr,
    corr, 1   , corr,
    corr, corr, 1   ), 3, 3)

# 1 1 1 ... 1 0 0 0 ... 0
x <- c(rep(1, n1fam), rep(1, n1fam), rep(1, n1fam),
      rep(1, n2fam), rep(0, n2fam), rep(0, n2fam))
# 1 2 ... n1fam 1 2 ... n1fam ...
id <- c(1:n1fam, 1:n1fam, 1:n1fam,
   (n1fam+1:n2fam), (n1fam+1:n2fam), (n1fam+1:n2fam))
power <- rep(0, numsim) # initialize vector for results
```

The concatenate function (`c()`) is used to glue together the appropriate elements of the design matrices and underlying correlation structure.

```
for (i in 1:numsim) {
   cat(i," ")
   # all three exposed
   grp1 <- mvrnorm(n1fam, c(effect, effect, effect), vmat)

   # only first exposed
   grp2 <- mvrnorm(n2fam, c(effect, 0,      0),      vmat)

   # concatenate the output vector
   y <- c(grp1[,1], grp1[,2], grp1[,3],
          grp2[,1], grp2[,2], grp2[,3])

   group <- groupedData(y ~ x | id)   # specify dependence structure
   res <- lme(group, random = ~ 1)    # fit random intercept model
   pval <- summary(res)$tTable[2,5]   # grab results for main parameter
   power[i] <- pval<=0.05             # is it statistically significant?
}

cat("\nEmpirical power for effect size of ", effect,
   " is ", round(sum(power)/numsim,3), ".\n", sep="")
cat("95% confidence interval is",
   round(prop.test(sum(power), numsim)$conf.int, 3), "\n")
```

This yields the following estimate.

```
Empirical power for effect size of 0.35 is 0.855.
95% confidence interval is 0.831 0.876
```

6.2 Generate data from generalized linear random effects model

In this example, we generate data from clustered data with a dichotomous outcome, an example of a generalized linear mixed model (4.2.6). In the code below, for 1500 clusters (denoted by `id`) there is a cluster invariant predictor (X_1), 3 observations within each cluster (denoted by X_2) and a linear effect of order within cluster, and an additional predictor which varies between clusters (X_3). The dichotomous outcome Y is generated from these predictors using a logistic link incorporating a random intercept for each cluster.

SAS

```
data sim;
   sigbsq=4; beta0=-2; beta1=1.5; beta2=0.5; beta3=-1; n=1500;
   do i = 1 to n;
      x1 = (i lt (n+1)/2);
      randint = normal(0) * sqrt(sigbsq);
      do x2 = 1 to 3 by 1;
         x3 = uniform(0);
         linpred = beta0 + beta1*x1 + beta2*x2 + beta3*x3 + randint;
         expit = exp(linpred)/(1 + exp(linpred));
         y = (uniform(0) lt expit);
         output;
      end;
   end;
run;
```

This model can be fit using `proc nlmixed` or `proc glimmix`, as shown below. For large datasets like this one, `proc nlmixed` (which uses numerical approximation to integration) can take a prohibitively long time to fit. On the other hand, `proc glimmix` can have trouble converging with the default maximization technique. We show options which use a maximization technique that may be helpful in such cases.

```
proc nlmixed data=sim qpoints=50;
   parms b0=1 b1=1 b2=1 b3=1;
   eta = b0 + b1*x1 + b2*x2 + b3*x3 + bi1;
   mu = exp(eta)/(1 + exp(eta));
   model y ~ binary(mu);
   random bi1 ~ normal(0, g11) subject=i;
   predict mu out=predmean;
run;
```

or

```
proc glimmix data=sim order=data;
   nloptions maxiter=100 technique=dbldog;
   model y = x1 x2 x3 / solution dist=bin;
   random int / subject=i;
run;
```

R

```
library(lme4)
n <- 1500; p <- 3; sigbsq <- 4
beta <- c(-2, 1.5, 0.5, -1)
id <- rep(1:n, each=p)    # 1 1 ... 1 2 2 ... 2 ... n
x1 <- as.numeric(id < (n+1)/2)  # 1 1 ... 1 0 0 ... 0
randint <- rep(rnorm(n, 0, sqrt(sigbsq)), each=p)
x2 <- rep(1:p, n)         # 1 2 ... p 1 2 ... p ...
x3 <- runif(p*n)
linpred <- beta[1] + beta[2]*x1 + beta[3]*x2 + beta[4]*x3 + randint
expit <- exp(linpred)/(1 + exp(linpred))
y <- runif(p*n) < expit

glmmres <- lmer(y ~ x1 + x2 + x3 + (1|id), family=binomial(link="logit"))
```

6.3 Generate correlated binary data

Correlated dichotomous outcomes Y_1 and Y_2 can be generated by finding the probabilities corresponding to the 2×2 table as a function of the marginal expectations and correlation using the methods of Lipsitz and colleagues [48]. Here we generate a sample of 1000 values where: $P(Y_1 = 1) = .15, P(Y_2 = 1) = .25$ and $\text{Corr}(Y_1, Y_2) = 0.40$.

SAS

```
data test;
   p1=.15; p2=.25; corr=0.4;
   p1p2=corr*sqrt(p1*(1-p1)*p2*(1-p2)) + p1*p2;
   do i = 1 to 10000;
      cat=rand('TABLE', 1-p1-p2+p1p2, p1-p1p2, p2-p1p2);
      y1=0;
      y2=0;
      if cat=2 then y1=1;
      else if cat=3 then y2=1;
      else if cat=4 then do;
         y1=1;
         y2=1;
      end;
      output;
   end;
run;
```

R

```
p1 <- .15; p2 <- .25; corr <- .4; n <- 10000
p1p2 <- corr*sqrt(p1*(1-p1)*p2*(1-p2)) + p1*p2
library(Hmisc)
vals <- rMultinom(matrix(c(1-p1-p2+p1p2, p1-p1p2, p2-p1p2, p1p2), 1, 4), n)
y1 <- rep(0, n); y2 <- rep(0, n)   # put zeroes everywhere
y1[vals==2 | vals==4] <- 1         # and replace them with ones
y2[vals==3 | vals==4] <- 1         # where needed
rm(vals, p1, p2, p1p2, corr, n)    # cleanup
```

The generated data is close to the desired values.

```
proc corr data=test;
   var y1;
   with y2;
run;

            The CORR Procedure
          1 With Variables:     y2
          1      Variables:     y1

 Variable       N      Mean    Std Dev      Sum   Minimum    Maximum
 y2         10000   0.25470    0.43571     2547         0    1.00000
 y1         10000   0.15290    0.35991     1529         0    1.00000

                             y1

                 y2      0.41107
                          <.0001
```

```
> cor(y1, y2)
[1] 0.3918081
> mean(y1)
[1] 0.1507
> mean(y2)
[1] 0.2484
```

6.4 Read variable format files and plot maps

Sometimes datasets are stored in variable format. For example, U.S. Census boundary files (available from `http://www.census.gov/geo/www/cob/index.html`) are available in both proprietary and ASCII formats. An example ASCII file describing the counties of Massachusetts is available on the book web site. The first few lines are reproduced here.

```
         1       -0.709816806854972E+02         0.427749187746914E+02
      -0.709148990000000E+02        0.428865890000000E+02
      -0.709148860000000E+02        0.428865640000000E+02
      -0.709148860000000E+02        0.428865640000000E+02
      -0.709027680000000E+02        0.428865300000000E+02
      -0.708861360000000E+02        0.428826100000000E+02
      -0.708837340828846E+02        0.428812223551543E+02
...
      -0.709148990000000E+02        0.428865890000000E+02
END
```

The first line contains an identifier for the county (linked with a county name in an additional file) and a latitude and longitude centroid within the polygon representing the county defined by the remaining points. The remaining points on the boundary do not contain the identifier. After the lines with the points, a line containing the word "END" is included. In addition, the county boundaries contain different numbers of points.

6.4.1 Read input files

Reading this kind of data requires some care in programming. We begin with SAS.

SAS

```
filename census1 url
"http://www.math.smith.edu/sasr/datasets/co25_d00.dat";

data pcts cents;
infile census1;
retain cntyid;
input @1 endind $3. @;   /* the trailing '@' means to hold onto this line */
if endind ne 'END' then do;
   input @7 neglat $1. @;   /* if this line does not say 'END', then
                               check to see if the 7th character is '-' */
   if neglat eq '-' then do;   /* if so, it has a boundary point */
     input @7 x y;
      output pcts;              /* write out to boundary dataset */
   end;
   else if neglat ne '-' then do;   /* if not, it must be the centroid */
      input @9 cntyid 2. x y ;
  output cents;                 /* write it to the centroid dataset */
   end;
end;
run;
```

Note: Two datasets are defined in the `data` statement, and explicit `output` statements are used to specify which lines are output to which datasets. The `@` designates the position on the line which is to be read, and also "holds" the line for further reading after the end of an `input` statement. The county names, which can be associated by the county identifier, are stored in another dataset.

```
filename census2 url
"http://www.math.smith.edu/sasr/datasets/co25_d00a.dat";

data cntynames;
infile census2 DSD;
   format cntyname $17. ;
   input cntyid 2. cntyname $;
run;
```

To get the names onto the map, we have to merge the centroid location dataset with the county names dataset. They have to be sorted first.

```
proc sort data=cntynames; by cntyid; run;
proc sort data=cents; by cntyid; run;
```

Note that in the preceding code we depart from the convention of requiring a new line for every statement; simple procedures like these are a convenient place to reduce the line length of the code.

In R we begin by reading in all of the input lines, keeping track of how many counties have been observed (based on how many lines include `END`). This information is needed for housekeeping purposes when collecting map points for each county.

R

```
# read in the data
input <- readLines("http://www.math.smith.edu/sasr/datasets/co25_d00.dat",
   n=-1)
# figure out how many counties, and how many entries
num <- length(grep("END", input))
allvals <- length(input)
numentries <- allvals-num
# create vectors to store data
county <- numeric(numentries); lat <- numeric(numentries)
long <- numeric(numentries)
```

```
curval <- 0   # number of counties seen so far
# loop through each line
for (i in 1:allvals) {
   if (input[i]=="END") {
      curval <- curval + 1
   } else {
      # remove extraneous spaces
      nospace <- gsub("[ ]+", " ", input[i])
      # remove space in first column
      nospace <- gsub("^ ", "", nospace)
      splitstring <- as.numeric(strsplit(nospace, " ")[[1]])
      len <- length(splitstring)
      if (len==3) {  # new county
         curcounty <- splitstring[1]; county[i-curval] <- curcounty
         lat[i-curval] <- splitstring[2]; long[i-curval] <- splitstring[3]
      } else if (len==2) { # continue current county
         county[i-curval] <- curcounty; lat[i-curval] <- splitstring[1]
         long[i-curval] <- splitstring[2]
      }
   }
}
```

Each line of the input file is processed in turn. The `strsplit()` function is used to parse the input file. Lines containing `END` require incrementing the count of counties seen to date. If the line indicates the start of a new county, the new county number is saved. If the line contains 2 fields (another set of latitudes and longitudes), then this information is stored in the appropriate index (`i-curval`) of the output vectors.

Next we read in a dataset of county names. Later we'll plot the Massachusetts counties, and annotate the plot with the names of the counties.

```
# read county names
countynames <-
   read.table("http://www.math.smith.edu/sasr/datasets/co25_d00a.dat",
   header=FALSE)
names(countynames) <- c("county", "countyname")
```

6.4.2 Plotting maps

In SAS, we're ready to merge the two datasets. At the same time, we'll include the variables needed by the `annotate` facility to put data from the dataset onto the map. The variables

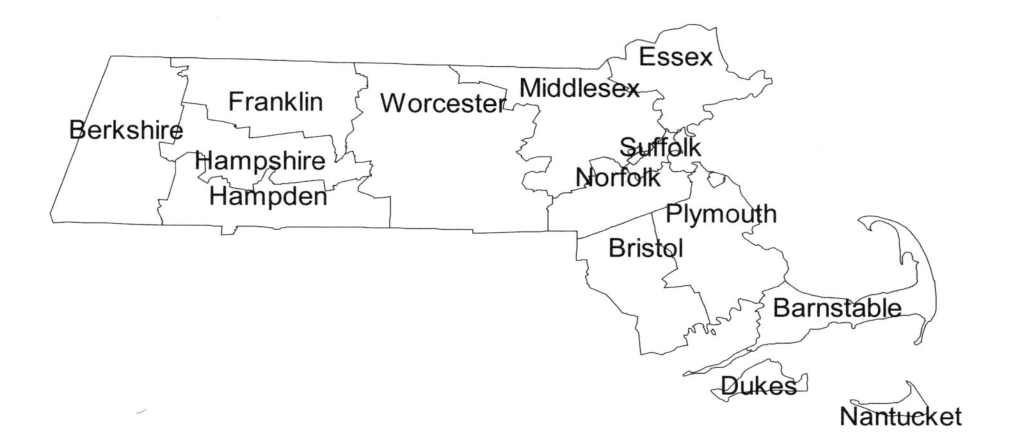

Figure 6.1: Massachusetts counties

`function`, `style`, `color`, `position`, `when`, `size`, and the `?sys` variables all describe aspects of the text to be placed onto the plot.

```
data nameloc;
   length function style color $ 8 position $ 1 text $ 20;
   retain xsys ysys "2" hsys "3" when "a";
merge cntynames cents;
by cntyid;
   function="label"; style="swiss"; text=cntyname; color="black";
   size=3; position="5";
   output;
run;
```

Finally, we can make the map. The `annotate` option (5.2) tells SAS to use the named dataset to mark up the map.

```
ods pdf file="map_plot.pdf";
pattern1 value=empty;
proc gmap map=pcts data=pcts;
   choro const / nolegend coutline=black annotate=nameloc;
   id cntyid;
run; quit;
ods pdf close;
```

The `pattern` statement can be used to control the fill colors when creating choropleth maps. Here we specify that no fill is needed.

Results are displayed in Figure 6.1.

To create the map in R, we begin by determining the plotting region, creating the plot of boundaries, then adding the county names at the internal point that was provided.

```
xvals <- c(min(lat), max(lat))
yvals <- c(range(long))
pdf("massachusettsmap.pdf")
plot(xvals, yvals, pch=" ", xlab="", ylab="", xaxt="n", yaxt="n")
counties <- unique(county)
for (i in 1:length(counties)) {
   # first element is an internal point
   polygon(lat[county==counties[i]][-1], long[county==counties[i]][-1])
   # plot name of county using internal point
   text(lat[county==counties[i]][1], long[county==counties[i]][1],
      countynames$countyname[i])
}
dev.off()
```

Since the first set of points is in the interior of the county, these are not included in the values given to the polygon function (see indexing, section B.4.2).

The `pdf()` function is used to create an external graphics file (see 5.4.1, creating PDF files). When all plotting commands are complete, the `dev.off()` function is used to close the graphics device.

The plots from SAS and R differ only with respect to the default font used, so we display the results only once, in Figure 6.1. Many other maps as well as more sophisticated projections are supported with the `maps` package (see also the CRAN Spatial Statistics Task View).

6.5 Missing data: multiple imputation

Missing data is ubiquitous in most real-world investigations. Here we demonstrate some of the capabilities for fitting incomplete data regression models using multiple imputation [68, 76, 33] implemented with chained equation models [92, 64].

In this example we replicate an analysis from section 4.6.1 in a version of the HELP dataset that includes missing values for several of the predictors. While not part of the regression model of interest, the `mcs` and `pcs` variables are included in the imputation models, which may make the missing at random assumption more plausible [13].

SAS

```
filename myurl url "http://www.math.smith.edu/sasr/datasets/helpmiss.csv"
   lrecl=704;

proc import replace datafile=myurl out=help dbms=dlm;
   delimiter=',';
   getnames=yes;
run;
```

```
ods select misspattern;
proc mi data=help nimpute=0;
   var homeless female i1 sexrisk indtot mcs pcs;
run;
ods select all;
```

In the SAS code above, we read the data and print a summary of the missing data patterns.

```
                          Missing Data Patterns

Group  homeless  female  i1  sexrisk  indtot  mcs  pcs       Freq

    1  X         X       X   X        X       X    X          454
    2  X         X       X   X        X       .    .            2
    3  X         X       X   X        .       X    X           13
    4  X         X       X   .        .       X    X            1
                        Missing Data Patterns
                  ----------------Group Means---------------
Group    Percent         homeless         female            i1

    1      96.60         0.462555       0.237885     17.920705
    2       0.43         1.000000              0     13.000000
    3       2.77         0.461538       0.230769     31.307692
    4       0.21         1.000000              0     13.000000
                        Missing Data Patterns
         ---------------------Group Means--------------------
Group          sexrisk         indtot            mcs            pcs

    1         4.638767      35.729075      31.662403      48.018233
    2         7.000000      35.500000              .              .
    3         4.153846              .      27.832265      49.931599
    4                .              .      28.452675      49.938469
```

Since the pattern of missingness is non-monotone, our options for imputing within SAS are somewhat limited. In the code below, we impute using MCMC. This is not strictly appropriate, since this technique assumes multivariate normal data, which is clearly not the case here. For a summary of multiple imputation options available in SAS, see [33]. An alternative would be to use IVEware, a free suite of SAS macros [65].

```
proc mi data=helpmiss nimpute=20 out=helpmi20 noprint;
   mcmc chain=multiple;
   var homeless female i1 sexrisk indtot mcs pcs;
run;
```

The output dataset `helpmi20` has 20 completed versions of the original dataset, along with an additional variable, `_imputation_`, which identifies the completed versions. We use the `by` statement in SAS to fit a logistic regression within each completed dataset.

```
ods select none;
ods output parameterestimates=helpmipe covb=helpmicovb;
proc logistic data=helpmi20 descending;
by _imputation_;
   model homeless=female i1 sexrisk indtot / covb;
run;
ods select all;
```

Note the use of the `ods select none` statement to suppress all printed output and to save the parameter estimates and their estimated covariance matrix for use in multiple imputation. The multiple imputation inference is performed in `proc mianalyze`.

```
proc mianalyze parms = helpmipe covb=helpmicovb;
   modeleffects intercept female i1 sexrisk indtot;
run;
```

This generates a fair amount of output; we reproduce only the parameter estimates and their standard errors.

```
                         Parameter Estimates

Parameter        Estimate        Std Error    95% Confidence Limits
intercept       -2.547100         0.596904     -3.71707      -1.37713
female          -0.241332         0.244084     -0.71973       0.23706
i1               0.023101         0.005612      0.01210       0.03410
sexrisk          0.057386         0.035842     -0.01286       0.12763
indtot           0.049641         0.015929      0.01842       0.08086
```

Within R, the Hmisc package includes many functions to describe missing data patterns as well as fit imputation models. We begin by reading in the data then using the na.pattern() function from the Hmisc library to characterize the patterns of missing values.

R

```
> ds <- read.csv("http://www.math.smith.edu/sasr/datasets/helpmiss.csv")
> smallds <- with(ds, data.frame(homeless, female, i1, sexrisk, indtot,
    mcs, pcs))
> summary(smallds)
    homeless           female             i1               sexrisk
 Min.   :0.000   Min.   :0.000   Min.   :  0.0   Min.   : 0.00
 1st Qu.:0.000   1st Qu.:0.000   1st Qu.:  3.0   1st Qu.: 3.00
 Median :0.000   Median :0.000   Median : 13.0   Median : 4.00
 Mean   :0.466   Mean   :0.236   Mean   : 18.3   Mean   : 4.64
 3rd Qu.:1.000   3rd Qu.:0.000   3rd Qu.: 26.0   3rd Qu.: 6.00
 Max.   :1.000   Max.   :1.000   Max.   :142.0   Max.   :14.00
                                                 NA's   : 1.00
     indtot           mcs              pcs
 Min.   : 4.0   Min.   : 6.76   Min.   :14.1
 1st Qu.:32.0   1st Qu.:21.66   1st Qu.:40.3
 Median :37.5   Median :28.56   Median :48.9
 Mean   :35.7   Mean   :31.55   Mean   :48.1
 3rd Qu.:41.0   3rd Qu.:40.64   3rd Qu.:57.0
 Max.   :45.0   Max.   :62.18   Max.   :74.8
 NA's   :14.0  NA's   : 2.00  NA's   : 2.0
> library(Hmisc)
> na.pattern(smallds)
pattern
0000000 0000011 0000100 0001100
    454       2      13       1
```

There are 14 subjects missing indtot, 2 missing mcs and pcs, and 1 missing sexrisk. In terms of patterns of missingness, there are 454 observations with complete data, 2 missing both mcs and pcs, 13 missing indtot alone, and 1 missing sexrisk and indtot. Fitting a logistic regression model using the available data (n=456) yields:

```
> glm(homeless ~ female + i1 + sexrisk + indtot, binomial, data=smallds)
Call:  glm(formula = homeless ~ female + i1 + sexrisk + indtot,
    family = binomial, data = smallds)

Coefficients:
(Intercept)       female           i1      sexrisk       indtot
    -2.5278      -0.2401       0.0232       0.0562       0.0493

(14 observations deleted due to missingness)
```

Next, the `mice()` function within the `mice` library is used to impute missing values for `sexrisk`, `indtot`, `mcs`, and `pcs`. These results are combined using `glm.mids()`, and results are pooled and reported. Note that by default, all variables within the `smallds` data frame are included in each of the chained equations (e.g., `mcs` and `pcs` are used as predictors in each of the imputation models).

```
> library(mice)
> imp <- mice(smallds, m=25, maxit=25, seed=42)
```

```
> summary(pool(glm.mids(homeless ~ female + i1 + sexrisk + indtot,
      family=binomial, data=imp)))
                est      se      t  df Pr(>|t|)   lo 95   hi 95
(Intercept) -2.5366 0.59460 -4.266 456 2.42e-05 -3.7050 -1.3681
female      -0.2437 0.24393 -0.999 464 3.18e-01 -0.7230  0.2357
i1           0.0231 0.00561  4.114 464 4.61e-05  0.0121  0.0341
sexrisk      0.0590 0.03581  1.647 463 1.00e-01 -0.0114  0.1294
indtot       0.0491 0.01582  3.105 455 2.02e-03  0.0180  0.0802
            missing     fmi
(Intercept)      NA 0.01478
female            0 0.00182
i1                0 0.00143
sexrisk           1 0.00451
indtot           14 0.01728
```

While the results are qualitatively similar, they do differ, which is not surprising given the different imputation models used.

6.6 Bayesian Poisson regression

Bayesian methods are increasingly commonly utilized, and implementations of many models are available within SAS as well as R. For SAS, the on-line documentation is a valuable resource: Contents; SAS Products; SAS/STAT; SAS/STAT User's guide, Introduction to Bayesian Analysis Procedures. For R, the CRAN Bayesian Inference Task View provides an overview of the packages that incorporate some aspect of Bayesian methodologies. In this example, we fit a Poisson regression model to the count of alcohol drinks in the HELP study as fit previously (4.6.2), this time using Markov Chain Monte Carlo methods. Specification of prior distributions is necessary for Bayesian analysis; diagnosis of convergence is a critical part of any MCMC model fitting (see Gelman et al., [26] for an accessible introduction).

SAS

```
proc import
   datafile='c:/book/help.csv'
   out=help dbms=dlm;
   delimiter=',';
   getnames=yes;
run;
```

```
proc genmod data=help;
   class substance;
   model i1 = female substance age / dist=poisson;
   bayes;
run;
```

```
                        Posterior Summaries

                                      Standard           Percentiles
Parameter              N      Mean   Deviation      25%       50%       75%
Intercept          10000    1.7774      0.0592   1.7366    1.7778    1.8179
female             10000   -0.1760      0.0279  -0.1946   -0.1761   -0.1570
substancealcohol   10000    1.1202      0.0345   1.0968    1.1201    1.1431
substancecocaine   10000    0.3026      0.0387   0.2764    0.3023    0.3283
age                10000    0.0132      0.0015   0.0122    0.0132    0.0142
```

```
                     Posterior Intervals

Parameter          Alpha     Equal-Tail Interval          HPD Interval
Intercept          0.050      1.6624      1.8925       1.6669      1.8957
female             0.050     -0.2317     -0.1221      -0.2294     -0.1201
substancealcohol   0.050      1.0531      1.1893       1.0484      1.1840
substancecocaine   0.050      0.2276      0.3795       0.2239      0.3755
age                0.050      0.0104      0.0161       0.0103      0.0161
```

Note: The `bayes` statement has options to control many aspects of the MCMC process; diagnostic graphics will be produced if an `ods graphics` statement is submitted. The above code produces the following posterior distribution characteristics for the parameters.

R

```
> ds <- read.csv("http://www.math.smith.edu/sasr/datasets/help.csv")
> attach(ds)
> library(MCMCpack)
Loading required package: coda
Loading required package: lattice
Loading required package: MASS
##
## Markov Chain Monte Carlo Package (MCMCpack)
## Copyright (C) 2003-2008 Andrew D. Martin, Kevin M. Quinn,
## and Jonh Hee Park
## Support provided by the U.S. National Science Foundation
## (Grants SES-0350646 and SES-0350613)
##
> posterior <- MCMCpoisson(i1 ~ female + as.factor(substance) + age)
The Metropolis acceptance rate for beta was 0.27891
```

```
> summary(posterior)
Iterations = 1001:11000
Thinning interval = 1
Number of chains = 1
Sample size per chain = 10000
1. Empirical mean and standard deviation for each variable,
   plus standard error of the mean:
                                Mean      SD Naive SE Time-series SE
(Intercept)                   2.8959 0.05963 5.96e-04       0.002858
female                       -0.1752 0.02778 2.78e-04       0.001085
as.factor(substance)cocaine -0.8176 0.02727 2.73e-04       0.001207
as.factor(substance)heroin  -1.1199 0.03430 3.43e-04       0.001333
age                           0.0133 0.00148 1.48e-05       0.000071

2. Quantiles for each variable:
                                2.5%     25%     50%     75%   97.5%
(Intercept)                   2.7807  2.8546  2.8952  2.9390  3.0157
female                       -0.2271 -0.1944 -0.1754 -0.1567 -0.1184
as.factor(substance)cocaine -0.8704 -0.8364 -0.8174 -0.7992 -0.7627
as.factor(substance)heroin  -1.1858 -1.1430 -1.1193 -1.0967 -1.0505
age                           0.0103  0.0122  0.0133  0.0143  0.0160
```

Note: Default plots are available for `MCMC` objects returned by `MCMCpack`. These can be displayed using the command `plot(posterior)`. Support for model assessment is provided in the `coda` (Convergence Diagnosis and Output Analysis) package.

6.7 Multivariate statistics and discriminant procedures

This section includes a sampling of commonly used multivariate, clustering methods and discriminant procedures [12, 82].

In SAS, summaries of these topics and how to implement related methods are discussed in the on-line help: Contents; SAS Products; SAS/STAT; SAS/STAT User's Guide under the headings "Introduction to Multivariate Procedures," "Introduction to Clustering Procedures," and "Introduction to Discriminant Procedures."

The Multivariate statistics, Cluster analysis and Psychometrics task views on CRAN provide additional descriptions of functionality available within R.

6.7.1 Cronbach's α

We begin by calculating Cronbach's α for the 20 items comprising the CESD (Center for Epidemiologic Studies–Depression scale).

```
ods select cronbachalpha;
proc corr data=help alpha nomiss;
   var f1a -- f1t;
run;
ods exclude none;

The CORR Procedure

 Cronbach Coefficient Alpha

Variables              Alpha
----------------------------
Raw                 0.760762
Standardized        0.764156
```

Note that the nomiss option is required in SAS to include only observations with all variables observed.

```
> library(multilevel)
> cronbach(cbind(f1a, f1b, f1c, f1d, f1e, f1f, f1g, f1h, f1i, f1j, f1k,
+    f1l, f1m, f1n, f1o, f1p, f1q, f1r, f1s, f1t))

$Alpha
[1] 0.761

$N
[1] 446
```

The observed α of 0.76 from the HELP study is relatively low: this may be due to ceiling effects for this sample of subjects recruited in a detoxification unit.

6.7.2 Factor analysis

Factor analysis is used to explain variability of a set of measures in terms of underlying unobservable factors. The observed measures can be expressed as linear combinations of the factors, plus random error. Factor analysis is often used as a way to guide the creation of summary scores from individual items. Here we consider a maximum likelihood factor analysis with varimax rotation for the individual items of the CESD (Center for Epidemiologic Studies–Depression) scale. The individual questions can be found in Table C.2, p. 279. We arbitrarily force three factors.

Before beginning in SAS, we exclude observations with missing values.

```
data helpcc;
set help;
        if n(of f1a--f1t) eq 20;
run;

ods select orthrotfactpat factor.rotatedsolution.finalcommunwgt;
proc factor data=helpcc nfactors=3 method=ml rotate=varimax;
   var f1a--f1t;
run;
```

```
The FACTOR Procedure
Rotation Method: Varimax

Final Communality Estimates and Variable Weights
Total Communality: Weighted = 15.332773   Unweighted = 7.811194

Variable    Communality       Weight
F1A          0.25549722   1.34316770
F1B          0.23225517   1.30252990
F1C          0.51565766   2.06467779
F1D          0.29270906   1.41401403
F1E          0.29893385   1.42636367
F1F          0.57894420   2.37499121
F1G          0.23471625   1.30675434
F1H          0.39897919   1.66400037
F1I          0.38389849   1.62312753
F1J          0.37453462   1.59881735
F1K          0.29461104   1.41765736
F1L          0.48551624   1.94346054
F1M          0.11832415   1.13419896
F1N          0.37735132   1.60602564
F1O          0.35641841   1.55382997
F1P          0.59280807   2.45558672
F1Q          0.28734113   1.40315708
F1R          0.53318869   2.14218252
F1S          0.72695038   3.66226205
F1T          0.47255864   1.89596701

Rotation Method: Varimax

              Factor1           Factor2           Factor3
F1A           0.44823          -0.19780           0.12436
F1B           0.42744          -0.18496           0.12385
F1C           0.61763          -0.29675           0.21479
F1D          -0.25073           0.45456          -0.15236
F1E           0.51814          -0.11387           0.13228
F1F           0.66562          -0.33478           0.15433
F1G           0.47079           0.03520           0.10880
F1H          -0.07422           0.62158          -0.08435
F1I           0.46243          -0.32461           0.25433
F1J           0.49539          -0.22585           0.27949
F1K           0.52291          -0.11535           0.08873
F1L          -0.27558           0.63987           0.01191
F1M           0.28394          -0.03699           0.19061
F1N           0.48453          -0.33040           0.18281
F1O           0.26188          -0.06977           0.53195
F1P          -0.07338           0.75511          -0.13125
F1Q           0.45736          -0.07107           0.27039
F1R           0.61412          -0.28168           0.27696
F1S           0.23592          -0.16627           0.80228
F1T           0.48914          -0.26872           0.40136
```

```
> res <- factanal(~ f1a + f1b + f1c + f1d + f1e + f1f + f1g + f1h +
+    f1i + f1j + f1k + f1l + f1m + f1n + f1o + f1p + f1q + f1r +
+    f1s + f1t, factors=3, rotation="varimax", na.action=na.omit,
+    scores="regression")
> print(res, cutoff=0.45, sort=TRUE)

Call:
factanal(x = ~f1a + f1b + f1c + f1d + f1e + f1f + f1g + f1h +
              f1i + f1j + f1k + f1l + f1m + f1n + f1o + f1p + f1q + f1r
+
              f1s + f1t, factors = 3, na.action = na.omit,
              scores = "regression",     rotation = "varimax")

Uniquenesses:
  f1a   f1b   f1c   f1d   f1e   f1f   f1g   f1h   f1i   f1j   f1k   f1l
0.745 0.768 0.484 0.707 0.701 0.421 0.765 0.601 0.616 0.625 0.705 0.514
  f1m   f1n   f1o   f1p   f1q   f1r   f1s   f1t
0.882 0.623 0.644 0.407 0.713 0.467 0.273 0.527

Loadings:
    Factor1 Factor2 Factor3
f1c  0.618
f1e  0.518
f1f  0.666
f1k  0.523
f1r  0.614
f1h         -0.621
f1l         -0.640
f1p         -0.755
f1o                  0.532
f1s                  0.802
f1a
f1b
f1d         -0.454
f1g  0.471
f1i  0.463
f1j  0.495
f1m
f1n  0.485
f1q  0.457
f1t  0.489

                Factor1 Factor2 Factor3
SS loadings       3.847   2.329   1.636
Proportion Var    0.192   0.116   0.082
Cumulative Var    0.192   0.309   0.391

Test of the hypothesis that 3 factors are sufficient.
The chi square statistic is 289 on 133 degrees of freedom.
The p-value is 1.56e-13
```

It is possible to interpret the item scores from the output. We see that the second factor loads on the reverse coded items (H, L, P, and D, see 1.13.3). Factor 3 loads on items O and S (*people were unfriendly* and *I felt that people dislike me*).

6.7.3 Recursive partitioning

Recursive partitioning is used to create a decision tree to classify observations from a dataset based on categorical predictors. Recursive partitioning is only available in SAS through SAS Enterprise Miner, a module not included with the educational license typically purchased by universities, or through relatively expensive third-party add-ons to SAS. Within R, this functionality is available within the `rpart` package. In this example, we attempt to classify subjects based on their homeless status, using gender, drinking, primary substance, RAB sexrisk, MCS, and PCS as predictors.

```
> library(rpart)
> sub <- as.factor(substance)
> homeless.rpart <- rpart(homeless ~ female + i1 + sub + sexrisk + mcs +
+      pcs, method="class", data=ds)
> printcp(homeless.rpart)

Classification tree:
rpart(formula = homeless ~ female + i1 + sub + sexrisk + mcs +
    pcs, data = ds, method = "class")

Variables actually used in tree construction:
[1] female  i1      mcs     pcs     sexrisk

Root node error: 209/453 = 0.5

n= 453

    CP nsplit rel error xerror xstd
1 0.10      0       1.0      1 0.05
2 0.05      1       0.9      1 0.05
3 0.03      4       0.8      1 0.05
4 0.02      5       0.7      1 0.05
5 0.01      7       0.7      1 0.05
6 0.01      9       0.7      1 0.05
```

Figure 6.7.3 displays the tree.

```
> plot(homeless.rpart)
> text(homeless.rpart)
```

To help interpret this model, we can assess the proportion of homeless among those with $i1 < 3.5$ by pcs divided at 31.94.

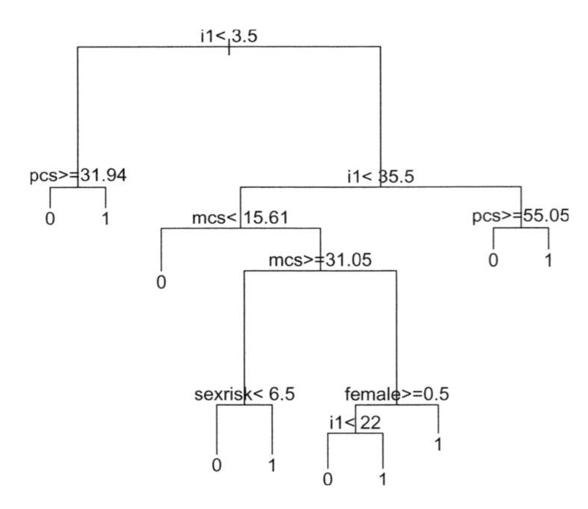

Figure 6.2: Recursive partitioning tree

```
> home <- homeless[i1<3.5]
> pcslow <- pcs[i1<3.5]<=31.94
> table(home, pcslow)

    pcslow
home FALSE TRUE
   0    89    2
   1    31    5

> rm(home, pcslow)
```

Amongst this subset, 71.4% (5 of 7) of those with low PCS scores are homeless, while only 25.8% (31 of 120) of those with PCS scores above the threshold are homeless.

6.7.4 Linear discriminant analysis

Linear (or Fisher) discriminant analysis is used to find linear combinations of variables that can separate classes. We use linear discriminant analysis to distinguish between homeless and non-homeless subjects, with a prior classification that half are in each group (default in SAS).

```
ods select lineardiscfunc classifiedresub errorresub;
proc discrim data=help out=ldaout;
   class homeless;
   var age cesd mcs pcs;
run;
```

```
The DISCRIM Procedure

Linear Discriminant Function for HOMELESS

Variable              0             1

Constant      -56.61467     -56.81613
AGE             0.76638       0.78563
CESD            0.86492       0.87231
MCS             0.68105       0.67569
PCS             0.74918       0.73750
```

```
Classification Summary for Calibration Data: WORK.HELP
Resubstitution Summary using Linear Discriminant Function

   From
HOMELESS          0            1          Total

       0        142          102            244
              58.20        41.80         100.00

       1         89          120            209
              42.58        57.42         100.00

   Total        231          222            453
              50.99        49.01         100.00

  Priors        0.5          0.5
```

```
Classification Summary for Calibration Data: WORK.HELP
Resubstitution Summary using Linear Discriminant Function

       Error Count Estimates for HOMELESS

                         0           1        Total

Rate                0.4180      0.4258       0.4219
Priors              0.5000      0.5000
```

```
> library(MASS)
> ngroups <- length(unique(homeless))
> ldamodel <- lda(homeless ~ age + cesd + mcs + pcs,
+    prior=rep(1/ngroups, ngroups))
> print(ldamodel)

Call:
lda(homeless ~ age + cesd + mcs + pcs, prior = rep(1/ngroups,
    ngroups))

Prior probabilities of groups:
  0   1
0.5 0.5

Group means:
   age cesd  mcs  pcs
0 35.0 31.8 32.5 49.0
1 36.4 34.0 30.7 46.9

Coefficients of linear discriminants:
         LD1
age   0.0702
cesd  0.0269
mcs  -0.0195
pcs  -0.0426
```

The results from SAS and R indicate that homeless subjects tend to be older, have higher CESD scores, and lower MCS and PCS scores.

Figure 6.3 displays the distribution of linear discriminant function values by homeless status; the discrimination ability appears to be slight. The distribution of the linear discriminant function values are shifted to the right for the homeless subjects, though there is considerable overlap between the groups.

```
axis1 label=("Prob(homeless eq 1)");

ods select "Histogram 1";
proc univariate data=ldaout;
   class homeless;
   var _1;
   histogram _1 / nmidpoints=20 haxis=axis1;
run;
```

```
> plot(ldamodel)
```

Details on display of `lda` objects can be found using `help(plot.lda)`.

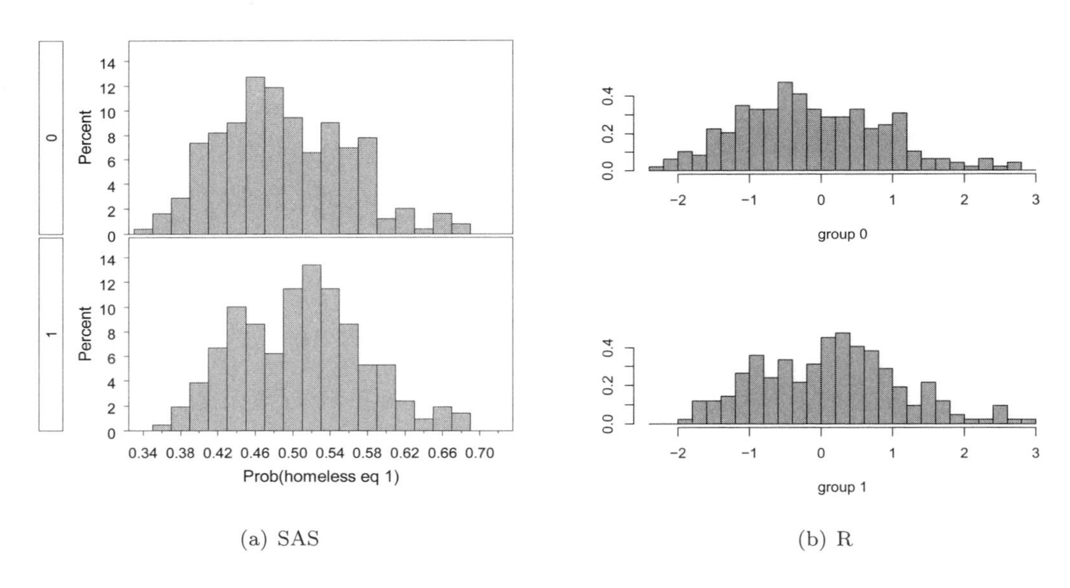

(a) SAS (b) R

Figure 6.3: Graphical display of assignment probabilities or score functions from linear discriminant analysis by actual homeless status

6.7.5 Hierarchical clustering

Many techniques exist for grouping similar variables or similar observations. These groups, or clusters, can be overlapping or disjoint, and are sometimes placed in a hierarchical structure so that some disjoint clusters share a higher-level cluster. Within SAS, the procedures `cluster`, `fastclus`, and `modeclus` can be used to find clusters of observations; the `varclus` and `factor` procedures can be used to find clusters of variables. The `tree` procedure can be used to plot tree diagrams from hierarchical clustering results. In R, there are many packages which perform clustering. Clustering tools included with the R distribution as part of the `stats` package include `hclust()` and `kmeans()`. The function `dendrogram()`, also in the `stats` package, plots tree diagrams. The `cluster()` package, included with the R distribution, contains functions `pam()`, `clara()`, and `diana()`. The CRAN Clustering Task View has more details. In this example, we cluster continuous variables from the HELP dataset.

```
ods exclude all;
proc varclus data=help outtree=treedisp centroid;
   var mcs pcs cesd i1 sexrisk;
run;
ods exclude none;

proc tree data=treedisp nclusters=5;
   height _varexp_;
run;
```

```
> cormat <- cor(cbind(mcs, pcs, cesd, i1, sexrisk),
+     use="pairwise.complete.obs")
> hclustobj <- hclust(dist(cormat))
```

Figure 6.4 displays the clustering. Not surprisingly, the MCS and PCS variables cluster together, since they both utilize similar questions and structures. The CESD and I1 variables cluster together, while there is a separate node for SEXRISK.

```
> plot(hclustobj)
```

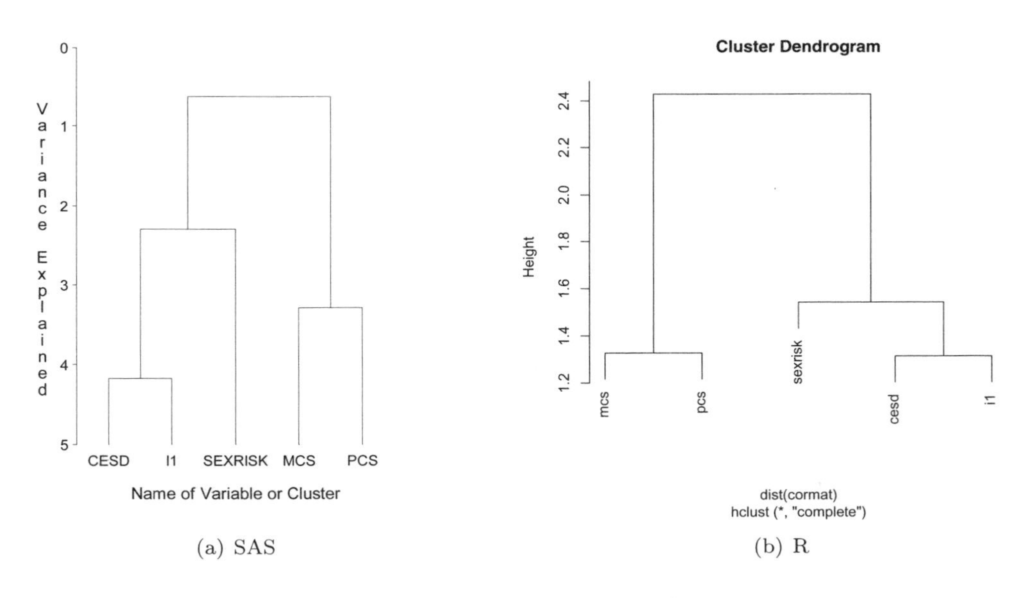

(a) SAS (b) R

Figure 6.4: Results from hierarchical clustering

6.8 Complex survey design

The appropriate analysis of sample surveys requires incorporation of complex design features, including stratification, clustering, weights, and finite population correction. These can be addressed in SAS and R for many common models. In this example, we assume that there are variables `psuvar` (cluster or PSU), `stratum` (stratification variable), and `wt` (sampling weight). Code examples are given to estimate the mean of a variable `x1` as well as a linear regression model.

SAS

```
proc surveymeans data=ds rate=fpcvar;
   cluster psuvar;
   strata stratum;
   weight wt;
   var x1 ... xk;
run;
```

or

```
proc surveyreg data=ds rate=fpcvar;
   cluster psuvar;
   strata stratum;
   weight wt;
   model y = x1 ... xk;
run;
```

Note: The `surveymeans` and `surveyreg` procedures account for complex survey designs with equivalent functionality to `means` and `reg`, respectively. Other survey procedures in SAS include `surveyfreq` and `surveylogistic`, which emulate procedures `freq` and `logistic`. The survey procedures share a `strata` statement to describe the stratification variables, a `weight` statement to describe the sampling weights, a `cluster` statement to specify the PSU or cluster, and a `rate` option (for the `proc` statement) to specify a finite population correction as a count or dataset. Additional options allow specification of the the total number of primary sampling units (PSUs) or a dataset with the number of PSUs in each stratum.

R

```
library(survey)
mydesign <- svydesign(id=~psuvar, strata=~stratum, weights=~wt,
   fpc=~fpcvar, data=ds)
meanres <- svymean(~ x1, mydesign)
regres <- svyglm(y ~ x1 + ... + xk, design=mydesign)
```

Note: The `survey` library includes support for many models. Illustrated above are means and linear regression models, with specification of PSU's, stratification, weight, and FPC.

6.9 Further resources

Rizzo's text [67] provides a comprehensive review of statistical computing tasks implemented using R, while [31] describes the use of R as a toolbox for mathematical statistics exploration.

Gelman, Carlin, Stern and Rubin [26] is an accessible introduction to Bayesian inference, while Albert [3] focuses on use of R for Bayesian computations.

Rubin's review [68] and Schafer's book [76] provide overviews of multiple imputation, while [92, 64] describe chained equation models. Review of software implementations of missing data models can be found in [34, 33].

Manly [12] and Tabachnick and Fidell [82] provide a comprehensive introduction to multivariate statistics. Särndal, Swensson, and Wretman [75] provides a readable overview of the analysis of data from complex surveys.

Appendix A

Introduction to SAS

The SAS™ system is a programming and data analysis package developed and marketed by SAS Institute, Cary NC. SAS markets many products; when installed together they result in an integrated environment. In this book we address software available in the Base SAS, SAS/STAT, SAS/GRAPH, SAS/ETS, and SAS/IML products. Base SAS provides a wide range of data management and analysis tools, while SAS/STAT and SAS/GRAPH provide support for more sophisticated statistics and graphics, respectively. We touch briefly on the IML (interactive matrix language) module, which provides extensive matrix functions and manipulation, and the ETS module, which supports time series tools and other specialized procedures. All of these products are typically included in educational institution installations, for which SAS Institute offers discounts.

SAS Institute also markets some products at reduced prices for individuals as well as for educational users. The "Learning Edition" lists at $US199 as of March, 2009, but limits its use to only 1,500 observations (rows in a dataset). More information can be found at `http://support.sas.com/learn/le/order.html`. Another option is SAS "OnDemand for Academics" (`http://www.sas.com/govedu/edu/programs/oda_account.html`) currently free for faculty and $60 for students. This option uses servers at SAS to run code and has a slightly more complex interface than the standard installation discussed in this book.

A.1 Installation

SAS products are available for a yearly license fee. Once licensed, a set of installation disks is mailed; this package includes detailed installation instructions tailored to the operating system for which the license was obtained. Also necessary is a special "setinit" file sent from SAS which functions as a password allowing installation of licensed products. An updated setinit file is sent upon purchase of a license renewal.

A.2 Running SAS and a sample session

Once installed, a recommended step for a new user is to start SAS and run a sample session. Starting SAS in a GUI environment opens a SAS window as displayed in Figure A.1.

The window is divided into two panes. On the left is a navigation pane with Results and Explorer tabs, while the right is an interactive windowing environment with Editor, Log, and Output Windows. Effectively, the right-hand pane is like a miniature graphical user interface (GUI) in itself. There are multiple windows, any one of which may be maximized, minimized, or closed. Their contents can also be saved to the operating system or printed.

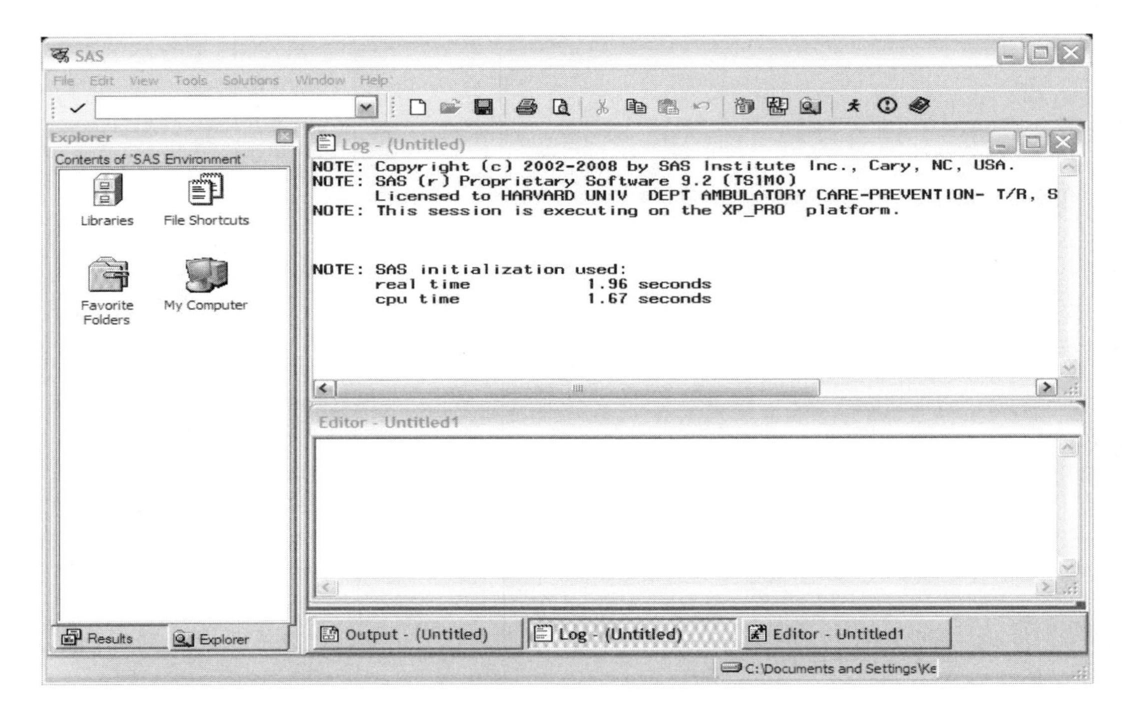

Figure A.1: SAS Windows interface

Depending on the code submitted, additional windows may open in this area. To open a window, click on its name at the bottom of the right-hand pane; to maximize or minimize within the SAS GUI, click on the standard icons your operating system uses for these actions.

On starting SAS, the cursor will appear in the Editor window. Commands such as those in the sample session which follows are typed there. They can also be read into the window from previously saved text files using *File; Open Program* from the menu bar. Typing the code doesn't do anything, even if there are carriage returns in it. To run code, it must be *submitted* to SAS; this is done by clicking the submit button in the GUI as in Figure A.2 or using keyboard shortcuts. After code is submitted SAS processes the code. Results are not displayed in the Editor window, but in the Output window, and comments from SAS on the commands which were run are displayed in the Log window. If output lines (typically analytic results) are generated, the Output window will jump to the front.

In the left-hand pane, the Explorer tab can be used to display datasets created within the current SAS session or found in the operating system. The datasets are displayed in a spreadsheet-like format. Navigation within the Explorer pane uses idioms familiar to users of GUI-based operating systems. The Results tab allows users to navigate among the output generated during the current SAS session. The Explorer and Results panes can each be helpful in reviewing data and results, respectively.

As a sample session, consider the following SAS code, which generates 100 normal variates (see 1.10.5) and 100 uniform variates (see 1.10.3), displays the first five of each (see 1.2.4), and calculates series of summary statistics (see 2.1.1). These commands would be typed directly into the Editor window:

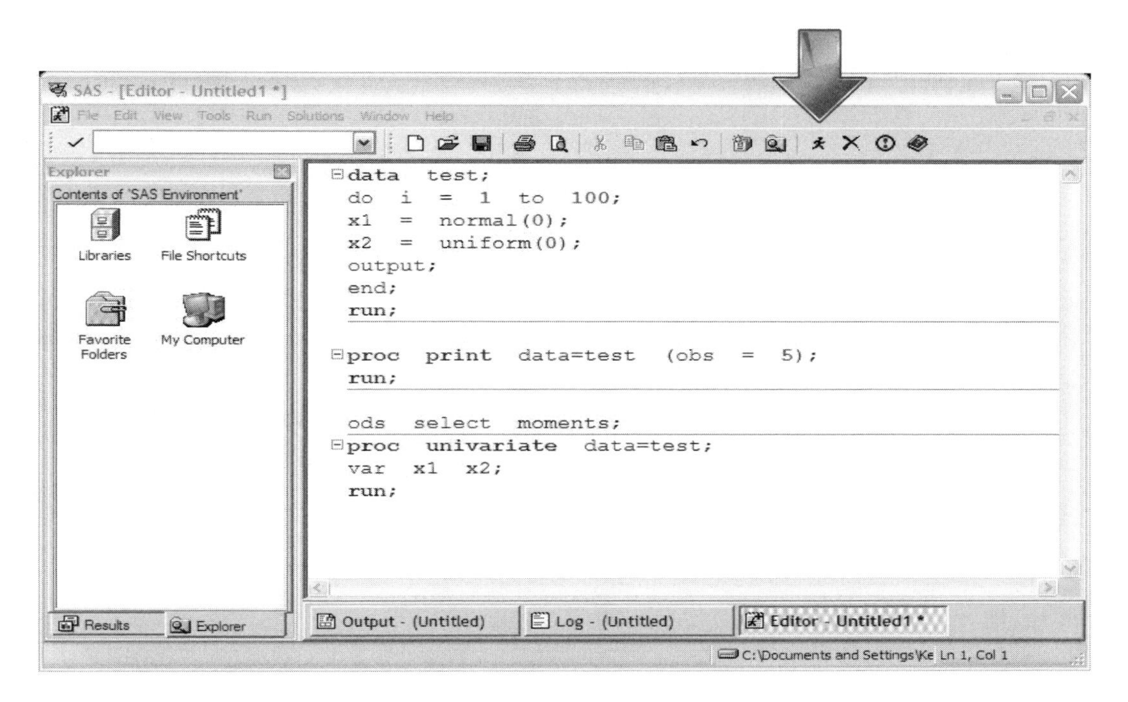

Figure A.2: Running a SAS program

```
data test;
   do i = 1 to 100;
      x1 = normal(0);
      x2 = uniform(0);
      output;
   end;
run;

proc print data=test (obs=5);
run;

ods select moments;
proc univariate data=test;
   var x1 x2;
run;
```

A user can run a section of code by selecting it using the mouse and clicking the "running figure" (submit) icon near the right end of the toolbar as shown in Figure A.2. Clicking the submit button when no text is selected will run all of the contents of the window. This code is available for download from the book website: `http://www.math.smith.edu/sasr/examples/sampsess.sas`.

We discuss each block of code in the example to highlight what is happening.

```
data test;
   do i = 1 to 100;
      x1 = normal(0);
      x2 = uniform(0);
      output;
   end;
run;
```

After selecting and submitting the above code the Output window will be empty, since no output was requested, but the log window will contain some new information:

```
1   data test;
2      do i = 1 to 100;
3         x1 = normal(0);
4         x2 = uniform(0);
5         output;
6      end;
7   run;

NOTE: The dataset WORK.TEST has 100 observations and 3 variables.
NOTE: DATA statement used (Total process time):
      real time             0.01 seconds
      cpu time              0.01 seconds
```

This indicates that the commands ran without incident, creating a dataset called `WORK.TEST` with 100 rows and three columns (one for `i`, one for `x1`, and one for `x2`). The line numbers can be used in debugging code.

Next consider the `proc print` code.

```
proc print data=test (obs=5);
run;
```

When these commands are submitted, SAS will generate the following in the Output window. Note that only 5 observations are shown because `obs=5` was specified (A.6.1). Omitting it will cause all 100 lines of data to be printed.

```
Obs    i        x1           x2

  1    1     0.38741     0.72843
  2    2     0.73014     0.37995
  3    3     1.48292     0.85374
  4    4    -1.86685     0.87779
  5    5    -0.33795     0.20864
```

Finally, data are summarized by submitting the lines specifying the `univariate` procedure.

```
ods select moments;
proc univariate data=test;
   var x1 x2;
run;
ods select all;
```

```
The UNIVARIATE Procedure
Variable:  x1

N                         100     Sum Weights               100
Mean               0.01009023     Sum Observations   1.00902341
Std Deviation      0.92945544     Variance           0.86388742
Skewness           -0.6296919     Kurtosis           -0.0743347
Uncorrected SS     85.5350355     Corrected SS       85.5248542
Coeff Variation    9211.43585     Std Error Mean     0.09294554
Variable:  x2

N                         100     Sum Weights               100
Mean               0.50565198     Sum Observations    50.565198
Std Deviation      0.29770927     Variance           0.08863081
Skewness           -0.0719154     Kurtosis           -1.1537095
Uncorrected SS     34.3428424     Corrected SS       8.77444994
Coeff Variation    58.8763178     Std Error Mean     0.02977093
```

As with the `obs=5` specified in the `proc print` statement above, the `ods select moments` statement causes a subset of the default output to be printed. By default, SAS often generates voluminous output that can be hard for new users to digest and would take up many pages of a book. We use the `ODS` system (A.7) to select pieces of the output throughout the book.

For each of these submissions, additional information is presented in the Log window. While some users may ignore the Log window unless the code did not work as desired, it is always a good practice to examine the log carefully, as it contains warnings about unexpected behavior as well as descriptions of errors which cause the code to execute incorrectly or not at all.

Note that the contents of the Editor, Log, and Output windows can be saved in typical GUI fashion by bringing the window to the front and using *File; Save* through the menus.

Figure A.3 shows the appearance of the SAS window after running the sample program. The Output window can be scrolled through to find results, or the Results tab shown in the left-hand pane can be used to find particular pieces of output more quickly. Figure A.4 shows the view of the dataset found through the Explorer window by clicking through *Libraries; Work; Test.* Datasets not assigned to permanent storage in the operating system (see writing native files, 1.2.1) are kept in a temporary library called the "Work" library.

A.3 Learning SAS and getting help

There are numerous tools available for learning SAS, of which at least two are built into the program. Under the Help menu in the Menu bar are "Getting Started with SAS Software" and "Learning SAS Programming." In the on-line help, under the `Contents` tab is "Learning to Use SAS" with many entries included. For those interested in learning about SAS but without access to a working version, some internet options include the excellent UCLA statistics website, which includes the "SAS Starter Kit" (`http://www.ats.ucla.edu/stat/sas/sk/default.htm`). While dated, the slide show available from the Oregon State University Statistics department could be useful (see `http://oregonstate.edu/dept/statistics/software/sas/2002seminar/index.htm`). SAS Institute offers several ways to get help. The central place to start is their web site where the front page for support is `http://support.sas.com/techsup`, which has links to discussion forums, support documents, and instructions for submitting an e-mail or phone request for technical support.

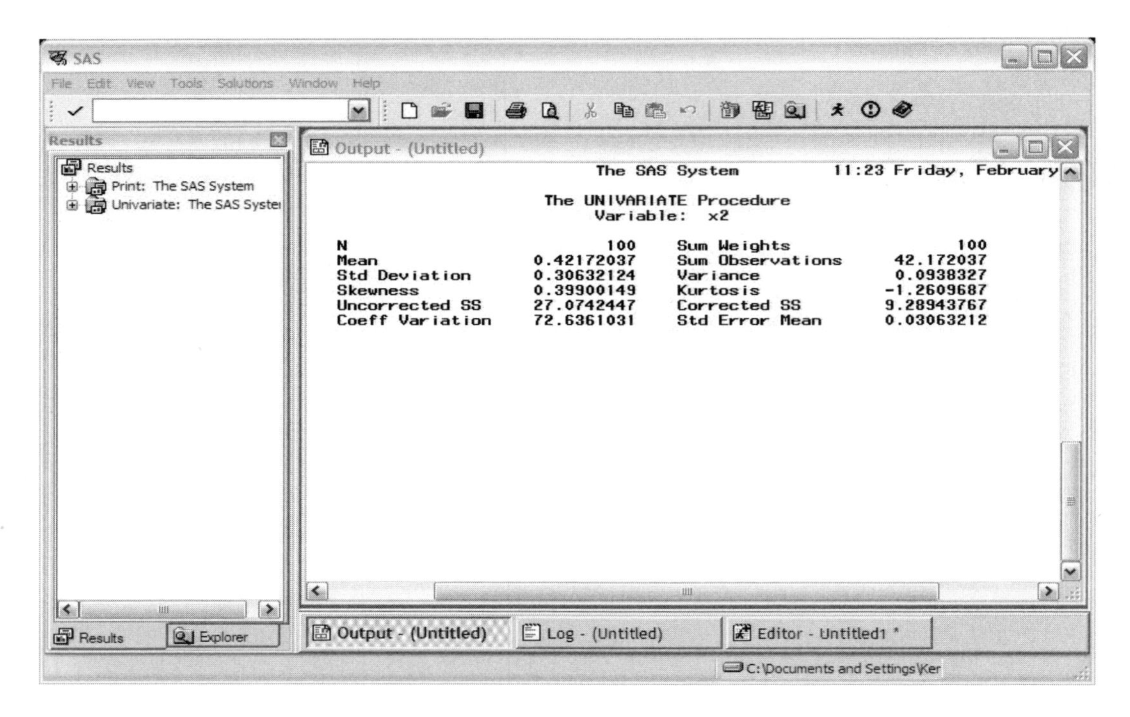

Figure A.3: The SAS window after running the sample session code

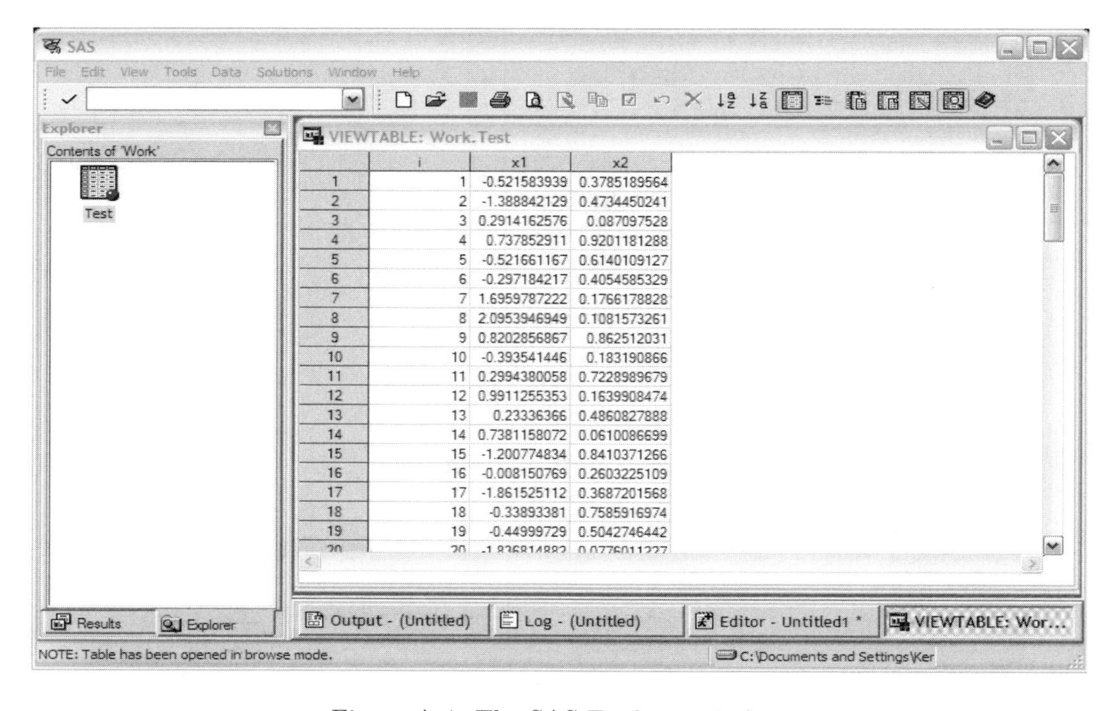

Figure A.4: The SAS Explorer window

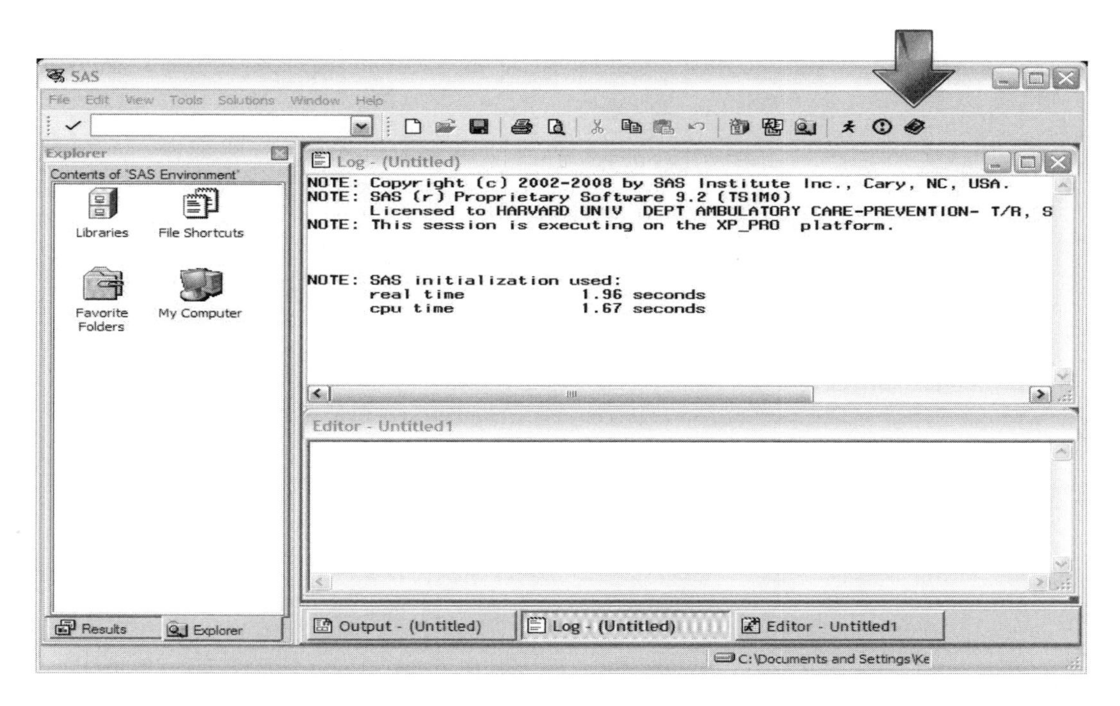

Figure A.5: Opening the on-line help

Complete documentation is included with SAS installation by default. Clicking the icon of the book with a question mark in the GUI (Figure A.5) will open a new window with a tool for viewing the documentation (Figure A.6). While there are `Contents, Index, Search`, and `Favorites` tabs in the help tool, we generally use the `Contents` tab as a starting point. Expanding the `SAS Products` folder here will open a list of SAS packages (i.e. Base SAS, SAS/STAT, etc.). Detailed documentation for the desired procedure can be found under the package which provides access to that `proc` or, as of SAS 9.2, in the alphabetical list of procedures found in: *Contents; SAS Products; SAS Procedures*. In the text, we provide occasional pointers to the on-line help, using the folder structure of the help tool to provide directions to these documents. Our pointers use the SAS 9.2 structure; earlier versions have a similar structure except that procedures must be located through their module. For example, to find the `proc mixed` documentation in SAS 9.2, you can use: Contents; SAS Products; SAS Procedures; MIXED, while before version 9.2, you would navigate to: Contents; SAS Products; SAS/STAT; SAS/STAT User's Guide; The MIXED Procedure.

A.4 Fundamental structures: data step, procedures, and global statements

Use of SAS can be broken into three main parts: the data step, procedures, and global statements. The data step is used to manage and manipulate data. Procedures are generally ways to do some kind of analysis and get results. Users of SAS refer to procedures as "procs." Global statements are generally used to set parameters and make optional choices that apply to the output of one or more procedures.

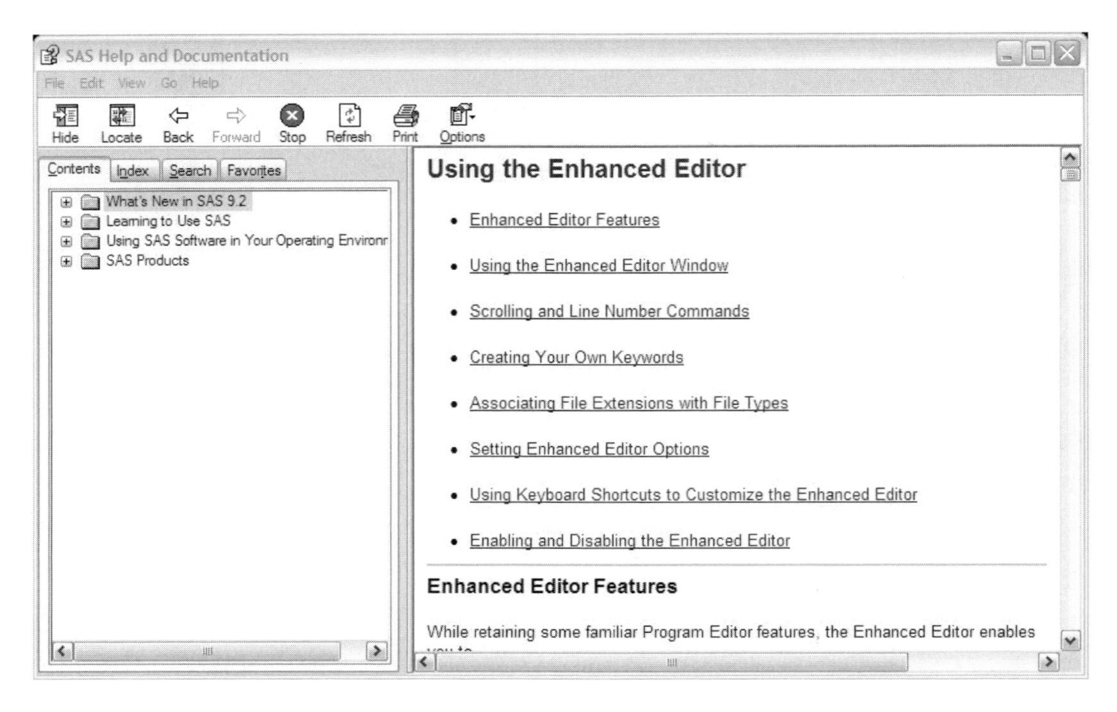

Figure A.6: The SAS Help and Documentation window

A typical data step might read as follows.

```
data newtest;
set test;
   logx = log(x);
run;
```

In this code a new variable named `logx` is created by taking the natural log of the variable `x`. The `data` step works by applying the instructions listed, sequentially, to each line of the dataset named using the `set` statement, then writing that line of data out to the dataset named in the `data` statement. Data steps and procedures are typically multi-statement collections. Both are terminated with a `run` statement. As shown above, statements in SAS are separated by semicolons, meaning that carriage returns and line breaks are ignored. When SAS reads the `run` statement in the example (when it reaches the "`;`" after the word `run`), it writes out the processed line of data, then repeats for each line of data until it reaches the end of the `set` dataset. In this example, a line of data is read from the `test` dataset, the `logx` variable is generated, and the line of data (including `logx`, `x`, and any other data stored in `test`) is written to the new dataset `newtest`.

A typical procedure in SAS might read as follows.

```
proc glm data=newtest;
   model y = logx / solution;
run;
```

Many procedures require multiple statements to function. For example, the `glm` procedure requires both a `proc glm` statement and a `model` statement.

Here, we show the two ways that *options* can be specified in SAS. One way is by simply listing optional syntax after the statement name. In the `proc glm` (3.1.1) statement above, we specify, using the `data` option, that the dataset that should be used is the `newtest` dataset. Without this option SAS defaults to using the most recently created dataset. As a matter of style, we always specify the dataset using the `data` option, which can be used with any and all `procs`. Naming datasets explicitly in each procedure minimizes errors and makes code clearer.

The `model` statement shown demonstrates another way that options are specified, namely after a forward slash. In general, this syntax is used when the main body of the statement may include separate words. For example, the slash in the `model` statement above separates the model specification from the options (here the `solution` option requests the parameter estimates in addition to the default ANOVA table).

We refer to any SAS code appearing between semicolons generically as "statements." Most statements appear within data steps or procs. Global statements are special statements that need not appear within a data step or a proc. An example would be the following code.

```
options ls=78 ps=60 nocenter;
```

This `options` statement affects the formatting of output pages, limiting the line length to 78 characters per line for 60 lines per page, while removing the default centering.

A.5 Work process: the cognitive style of SAS

A typical SAS work session involves first writing a `data` step or loading a saved command file (conventionally saved with a `.sas` extension) which might read in or perhaps modify a saved dataset. Then a `proc` is written to perform a desired analysis. The output is examined, and based on the results, the `data` step is modified to generate new variables, the `proc` is edited to choose new options, new `procs` are written, or some subset of these steps is repeated. At the end of the session, the dataset might be saved in the native SAS format, the commands saved in text format, and the results printed onto paper or saved (conventionally with a `.lst` extension).

A.6 Useful SAS background

A.6.1 Data set options

In addition to `data` steps for manipulating data, SAS allows on-the-fly modification of datasets. This approach, while less than ideal for documentation, can be a useful way to reduce code length: rather than create a new dataset with a subset of observations, or with a renamed variable, this can be done simultaneously with specifying the dataset to be used in a procedure. The syntax for these commands, called "data set options" in SAS documentation, is to list them in parentheses after naming the dataset. So, for example, to exclude extraneous variables in a dataset from an analysis dataset, the following code could be used to save time if the dataset were large.

```
proc ttest data=test2 (keep=x y);
   class x;
   var y;
run;
```

Another useful data set option limits the number of observations used from the named dataset.

```
proc ttest data=test2 (obs=60);
   class x;
   var y;
run;
```

A full list of data set options can be found in the on-line documentation: Contents; SAS Products; Base SAS; SAS 9.2 Language Reference: Dictionary; Dictionary of Language Elements; SAS Data Set Options.

A.6.2 Repeating commands for subgroups

A common problem in data analysis involves repeating some process for strata defined by a categorical variable. For example, a model might be needed for males and females separately, or for several different age groups. SAS provides an easy way to do this via the `sort` procedure and the `by` statement. Here we demonstrate using variables from the HELP dataset, assuming it has been read in using one of the methods described in section 1.1 and demonstrated at the outset of each example section.

```
proc sort data=ds;
   by female;
run;

proc glm data=ds;
   by female:
   model mcs = pcs;
run;
```

The `proc glm` code will generate regression output for each value of `female`. Many procedures support a `by` statement in this fashion. If the data have not been `sorted` previously, an error is likely.

A.6.3 Subsetting

It is often convenient to restrict the membership in a dataset or run analyses on a subset of observations. There are three main ways we do this in SAS. One is through the use of a subsetting `if` statement in a data step. The syntax for this is simply

```
data ...;
set ...;
   if condition;
run;
```

where condition is a logical statement such as `x eq 2` (see 1.11.2 for a discussion of logical operators). This includes only observations for which the condition is true, because when an `if` statement (1.11.2) does not include a `then`, the implied `then` clause is interpreted as "then output this line to the dataset; otherwise do not output it."

A second approach is a `where` statement. This can be used in a `data` step or in a procedure, and has a similar syntax.

```
proc ... data=ds;
   where condition;
...
run;
```

Finally, there is also a `where` data set option which can be used in a `data` step or a procedure; the syntax here is slightly different.

```
proc ... data=ds (where=(condition));
...
run;
```

The differences between the `where` statement and the `where` data set option are subtle and beyond our scope here. However, it is generally computationally cheaper to use a `where` approach than a subsetting `if`.

A.6.4 Formats and informats

SAS provides special tools for displaying variables or reading them in when they have complicated or unusual constructions in raw text. A good example for this is dates, for which June 27, 2009 might be written as, for example, `6-27-09, 27-6-09, 06/27/2009`, and so on. SAS stores dates as the integer number of days since December 31, 1959. To convert one of the aforementioned expressions to the desired storage value, `17710`, you use an *informat* to describe the way the data is written. For example, if the data were stored as the above expressions, you would use the informats `mmddyy8.`, `ddmmyy8.`, and `mmddyy10.` respectively to read them correctly as `17710`. An example of reading in dates is shown in section 1.1.2. More information on informats can be found in the on-line documentation: Contents; SAS Products; Base SAS; SAS 9.2 Language Reference: Dictionary; Informats.

In contrast, displaying data in styles other than that in which it is stored is done using the `informat`'s inverse, the `format`. The format for display can be specified within a `proc`. For example, if we plan a time series plot of `x*time` and want the x-axis labeled in quarters (i.e., `2010Q3`), we could use the following code, where the `time` variable is the integer-valued date. Information on formats can be found in the on-line documentation: Contents; SAS Products; Base SAS; SAS 9.2 Language Reference: Dictionary; Formats.

```
proc gplot data=ds;
   plot x*time;
   format time yyq6.;
run;
```

Another example is deciding how many decimal digits to display. For example, if you want to display 2 decimal places for variable `p` and 3 for variable `x`, you could use the following code.

```
proc print data=ds;
   var p x;
   format p 4.2 x 5.3;
run;
```

This topic is also discussed in 1.2.4.

A.7 Accessing and controlling SAS output: the Output Delivery System

Unlike many R commands, SAS does not provide access to most of the internal objects used in calculating results. Instead, it provides specific access to many objects of interest through various procedure statements. The ways to find these objects is idiosyncratic, and we have tried to highlight the most commonly needed objects in the text. This situation is roughly equivalent to the need in R to know the full name of an object before it can be accessed.

A much more general way to access and control output within SAS is through the `output delivery system` or (redundantly, as in "ATM machine") the `ODS` system. This is a very powerful and flexible system for accessing procedure results and controlling printed output. We use the `ODS` system mainly for two tasks: 1) to save procedure output into explicitly named datasets and 2) to suppress some printed output from procedures which generate lengthy output. In addition, we discuss using the `ODS` system to save output in useful file formats such as portable document format (PDF), hypertext markup language (HTML) or rich text format (RTF). Finally, we discuss `ODS` graphics, which add graphics to procedures' text output. We note that ODS has other uses beyond the scope of this book, and encourage readers to spend time familiarizing themselves with it.

A.7.1 Saving output as datasets and controlling output

Using ODS to save output or control the printed results involves two steps; first, finding out the name by which the `ODS` system refers to the output, and second, requesting that the dataset be saved as an ordinary SAS dataset or including or excluding it as output. The names used by the `ODS` system can be most easily found by running an `ods trace / listing` statement (later reversed using an `ods trace off` statement). The ods `outputname` thus identified can be saved using an `ods output outputname=newname` statement. A piece of output can be excluded using an `ods exclude outputname1 outputname2 ... outputnamek` statement or all but desired pieces excluded using the `ods select outputname1 outputname2 ... outputnamek` statement. These statements are each included before the procedure code which generates the output concerned. The `exclude` and `select` statements can be reversed using an `ods exclude none` or `ods select all` statement.

For example, to save the result of the t-test performed by `proc ttest` (2.4.1), the following code would be used. First, generate some data for the test.

```
data test2;
   do i = 1 to 100;
      if i lt 51 then x=1;
         else x=0;
      y = normal(0) + x;
      output;
   end;
run;
```

Then, run the t-test, including the `ods trace on / listing` statement to learn the names used by the `ODS` system.

```
ods trace on / listing;
proc ttest data=test2;
   class x;
   var y;
run;
ods trace off;
```

which would result in the following output.

```
Variable:  y

Output Added:
-------------
Name:       Statistics
Label:      Statistics
Template:   Stat.TTest.Statistics
Path:       Ttest.y.Statistics
-------------

x              N       Mean    Std Dev    Std Err    Minimum    Maximum

0             50     0.1403     1.0236     0.1448    -2.2392     2.1914
1             50     0.8941     0.9653     0.1365    -1.5910     2.7505
Diff (1-2)          -0.7538     0.9949     0.1990
```

```
Variable:  y

Output Added:
-------------
Name:       ConfLimits
Label:      Confidence Limits
Template:   Stat.TTest.ConfLimits
Path:       Ttest.y.ConfLimits
-------------

x           Method              Mean      95% CL Mean         Std Dev

0                             0.1403     -0.1506   0.4312      1.0236
1                             0.8941      0.6198   1.1685      0.9653
Diff (1-2)  Pooled           -0.7538     -1.1487  -0.3590      0.9949
Diff (1-2)  Satterthwaite    -0.7538     -1.1487  -0.3590

x           Method            95% CL Std Dev

0                             0.8550   1.2755
1                             0.8064   1.2029
Diff (1-2)  Pooled            0.8730   1.1567
Diff (1-2)  Satterthwaite
```

```
Variable:  y

Output Added:
-------------
Name:       TTests
Label:      T-Tests
Template:   Stat.TTest.TTests
Path:       Ttest.y.TTests
-------------
```

```
Method            Variances          DF    t Value    Pr > |t|

Pooled            Equal              98      -3.79      0.0003
Satterthwaite     Unequal        97.665      -3.79      0.0003
```

```
Variable:  y

Output Added:
-------------
Name:        Equality
Label:       Equality of Variances
Template:    Stat.TTest.Equality
Path:        Ttest.y.Equality
-------------

              Equality of Variances

Method       Num DF    Den DF    F Value    Pr > F

Folded F         49        49       1.12    0.6833
```

Note that failing to issue the `ods trace off` command will result in continued annotation of every piece of output. Similarly, when using the `ods exclude` and `ods select` statements, it is good practice to conclude each procedure with an `ods select all` or `ods exclude none` statement so that later output will be printed.

The previous output shows that the t-test itself (including the tests assuming equal and unequal variances) appears in output which the `ODS` system calls `ttests`, so the following code demonstrates how the test can be saved into a new dataset. Here we assign the new dataset the name `appendixattest`.

```
ods output ttests=appendixattest;
proc ttest data=test2;
   class x;
   var y;
run;

proc print data=appendixattest;
run;
```

and the `proc print` code results in the following output.

```
Obs   Variable   Method          Variances    tValue       DF     Probt

 1       y       Pooled           Equal        -3.79       98    0.0003
 2       y       Satterthwaite    Unequal      -3.79   97.665    0.0003
```

To run the t-ttest and print only these results, the following code would be used.

```
ods select ttests;
proc ttest data=test2;
   class x;
   var y;
run;
ods select all;

Variable:  y

Method           Variances        DF    t Value    Pr > |t|

Pooled           Equal            98      -3.79      0.0003
Satterthwaite    Unequal      97.665      -3.79      0.0003
```

This application is especially useful when running simulations, as it allows the results of procedures to be easily stored for later analysis.

The foregoing barely scratches the surface of what is possible using ODS. For further information, refer to the on-line help: Contents; SAS Products; Base SAS; SAS 9.2 Output Delivery System User's Guide.

A.7.2 Output file types and ODS destinations

The other main use of the ODS system is to generate output in a variety of file types. By default, SAS output is printed in the `output` window in the internal GUI. When run in batch mode, or when saving the contents of the output window using the GUI, this output is saved as a plain text file with a `.lst` extension. The ODS system provides a way to save SAS output in a more attractive form. As discussed in section 5.4, procedure output and graphics can be saved to named output files by using commands of the following form.

```
ods destinationname file="filename.ext";
```

The valid `destinationnames` include `pdf`, `rtf`, `latex`, and others. SAS refers to these file types as "destinations." It is possible to have multiple destinations open at the same time. For destinations other than `listing` (the output window), the `destination` must be closed before the results can be seen. This is done using the

```
ods destinationname close;
```

statement. Note that the default `listing` destination can also be closed; if there are no output destinations open, no results can be seen.

A.7.3 ODS graphics

The ODS system also allows users to incorporate text and graphical output from a procedure in an integrated document. This is done by "turning on" ODS graphics using an `ods graphics on` statement (as demonstrated in section 4.6.8), and then accepting default graphics or requesting particular plots using a `plots=plotnames` option to the procedure statement, where the valid plot names vary by procedure.

Special note for UNIX users: To generate ODS Graphics output in UNIX batch jobs, you must set the DISPLAY system option before creating the output. To set the display, enter the following command in the shell.

```
export DISPLAY=<ip_address>:0 (Korn shell)

DISPLAY=<ip_address>:0
export DISPLAY         (Bourne shell)
```

```
setenv DISPLAY=<ip_address>:0 (C shell)
```

In the above, `ip_address` is the fully qualified domain name or IP address, or the name of a display. Usually, the IP address of the UNIX system where SAS is running would be used. If you do not set the DISPLAY variable, then you get an error message in the SAS log. Additional information for UNIX users can be found in the on-line help: Contents; Using SAS Software in Your Operating Environment; SAS 9.2 Companion for UNIX Environments.

A.8 The SAS Macro Facility: writing functions and passing values

A.8.1 Writing functions

Unlike R, SAS does not provide a simple means for constructing functions which can be integrated with other code. However, it does provide a text-replacement capacity called the SAS Macro Language which can simplify and shorten code. The language also includes looping capabilities. We demonstrate here a simple macro to change the predictor in a simple linear regression example.

```
%macro reg1 (pred=);
   proc reg data=ds;
      model y = &pred;
   run;
%mend reg1;
```

In this example, we define the new macro by name (`reg1`) and define a single parameter which will be passed in the macro call; this will be referred to as `pred` within the macro. To replace `pred` with the submitted value, we use `&pred`. Thus the macro will run `proc reg` (3.1.1) with whatever text is submitted as the predictor of the outcome `y`. This macro would be called as follows.

```
%reg1(pred=x1);
```

When the `%macro` statements and the `%reg1` statement are run, SAS sees the following.

```
proc reg data=ds;
   model y = x1;
run;
```

If four separate regressions were required, they could then be run in four statements.

```
%reg1(pred=x1);
%reg1(pred=x2);
%reg1(pred=x3);
%reg1(pred=x4);
```

As with the Output Delivery System, SAS macros are a much broader topic than can be fully explored here. For a full introduction to its uses and capabilities, see the on-line help: Contents; SAS Products; Base SAS; SAS 9.2 Macro Language: Reference.

A.8.2 SAS macro variables

SAS also includes what are known as *macro variables*. Unlike SAS macros, macro variables are values that exist during SAS runs and are not stored within datasets. In general, a macro variable is defined with a `%let` statement.

```
%let macrovar=chars;
```

Note that the `%let` statement need not appear within a `data` step; it is a global statement. The value is stored as a string of characters, and can be referred to as `¯ovar`.

```
data ds;
   newvar=&macrovar;
run;
```

or

```
title "This is the &macrovar";
```

In the above example, the double quotes in the `title` statement allow the text within to be processed for the presence of macro variables. Enclosing the title text in single quotes will result in `¯ovar` appearing in the title, while the code above will replace `¯ovar` with the value of the `macrovar` macro variable.

While this basic application of macro variables is occasionally useful in coding, a more powerful application is to generate the macro variables within a SAS `data` step. This can be done using a `call symput` function as shown in 2.6.4.

```
data _null_;
...
   call symput('macrovar', x);
run;
```

This makes a new macro variable named `macrovar` which has the value of the variable `x`. The `_null_` dataset is a special SAS dataset which is not saved. It is efficient to use it when there is no need for a stored dataset.

A.9 Miscellanea

Official documentation provided by SAS refers to, for example `PROC GLM`. However, SAS is not case sensitive, with few exceptions. In this text we use lower case throughout. We find lower case easier to read, and prefer the ease of typing (both for coding and book composition) in lower case.

Since statements are separated by semicolons, multiple statements may appear on one line and statements may span lines. We usually type one statement per line in the text (and in practice), however. This prevents statements being overlooked among others appearing in the same line. In addition, we indent statements within a data step or proc, to clarify the grouping of related commands.

We prefer the fine control available through text-based commands. However, some people may prefer a point-and-click interface to the analytic tools available. SAS provides the SAS/Analyst application for such an approach; more information can be found at `http://support.sas.com/rnd/app/da/analyst/overview.html`. Another product is SAS/LAB; see `http://www.sas.com/products/lab` for more information on this. A third option is SAS/INSIGHT (`http://support.sas.com/documentation/onlinedoc/insight/ index`). These options may already be available in your installation.

SAS includes both `run` and `quit` statements. The `run` statement tells SAS to act on the code submitted since the most recent `run` statement (or since startup, if there has been no `run` statement submitted thus far). Some procedures allow multiple steps within the procedure without having to end it; key procedures which allow this are `proc gplot` and `proc reg`. This might be useful for model fitting and diagnostics with particularly large datasets in `proc reg`. In general, we find it a nuisance in graphics procedures, because the graphics are sometimes not actually drawn until the `quit` statement is entered. In the examples, we use the `run` statement in general and the `quit` statement when necessary, without further comment.

We find the SAS GUI to be a comfortable work environment and an aid to productivity. However, SAS can be easily run in batch mode. To use SAS this way, compose code in the text editor of your choice. Save the file (a `.sas` extension would be appropriate), then find it in the operating system. In Windows, a right-click on the file will bring up a list of potential actions, one of which is "Batch Submit with SAS 9.2." If this option is selected, SAS will run the file without opening the GUI. The output will be saved in the same directory with the same name but with a `.lst` extension; the log will be saved in the same directory with the same name but with a `.log` extension. Both output files are plain text files.

Appendix B

Introduction to R

This chapter provides a (brief) introduction to R, a powerful and extensible free software environment for statistical computing and graphics [38, 63]. The chapter includes a short history, installation information, a sample session, background on fundamental structures and actions, information about help and documentation, and other important topics.

R is a general purpose package that includes support for a wide variety of modern statistical and graphical methods (many of which are included through user contributed packages). It is available for most UNIX platforms, Windows and MacOS. R is part of the GNU project, and is distributed under a free software copyleft (`http://www.gnu.org/copyleft/gpl.html`). The R Foundation for Statistical Computing holds and administers the copyright of R software and documentation.

The first versions of R were written by Ross Ihaka and Robert Gentleman at the University of Auckland, New Zealand, while current development is coordinated by the R Development Core Team, a committed group of volunteers. As of January 2009 this consisted of Douglas Bates, John Chambers, Peter Dalgaard, Seth Falcon, Robert Gentleman, Kurt Hornik, Stefano Iacus, Ross Ihaka, Friedrich Leisch, Thomas Lumley, Martin Maechler, Duncan Murdoch, Paul Murrell, Martyn Plummer, Brian Ripley, Deepayan Sarkar, Duncan Temple Lang, Luke Tierney, and Simon Urbanek. Many hundreds of other people have contributed to the development of R or developed add-on libraries and packages on a volunteer basis.

R is similar to the S language, a flexible and extensible statistical environment originally developed in the 1980's at AT&T Bell Labs (now Lucent Technologies). Insightful Corporation has continued the development of S in their commercial software package S-PLUS™.

New users are encouraged to download and install R from the Comprehensive R archive network (CRAN, section B.1), complete the sample session in the Appendix of the *Introduction to R* document, also available from CRAN (see section B.2), then review this chapter.

B.1 Installation

The home page for the R project, located at `http://r-project.org`, is the best starting place for information about the software. It includes links to CRAN, which features precompiled binaries as well as source code for R, add-on packages, documentation (including manuals, frequently asked questions, and the R newsletter) as well as general background information. Mirrored CRAN sites with identical copies of these files exist all around the world. Updates are regularly posted on CRAN, which must be downloaded and installed.

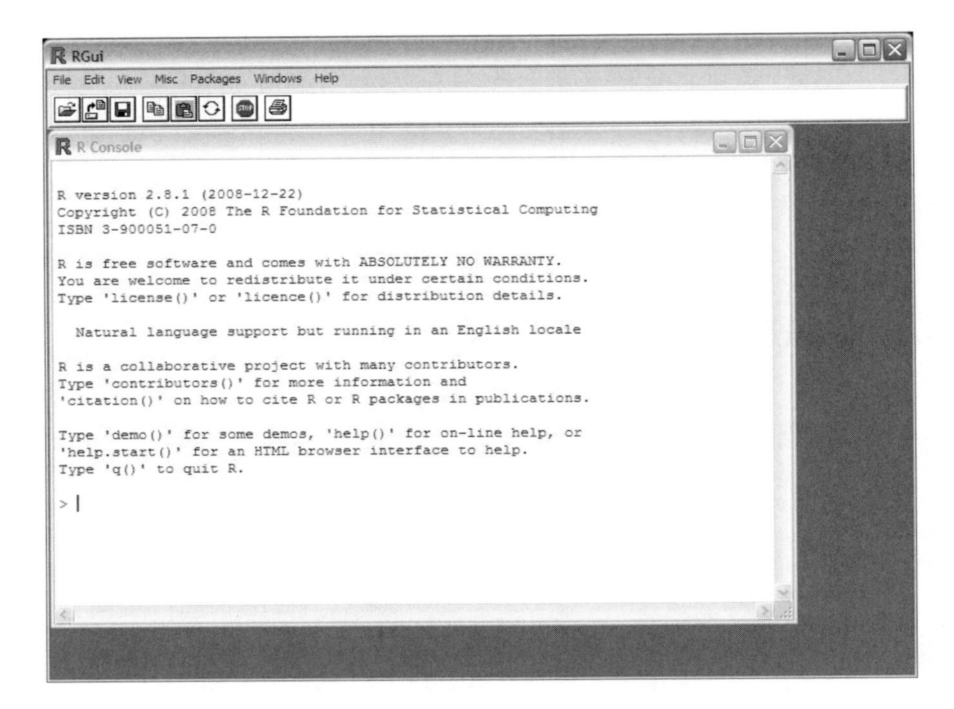

Figure B.1: R Windows graphical user interface

B.1.1 Installation under Windows

Pre-compiled distributions of R for Windows 95, 98, ME, NT4, 2000, XP, 2003 Server and Vista are available at CRAN. Two versions of the executable are available: `Rgui.exe`, which launches a self-contained windowing system that includes a command-line interface, and `Rterm.exe` which is suitable for batch or command-line use. A screenshot of the R graphical user interface (GUI) can be found in Figure B.1. More information on Windows specific issues can be found in the CRAN *R for Windows FAQ* (`http://cran.r-project.org/bin/windows/base/rw-FAQ.html`).

B.1.2 Installation under Mac OS X

A pre-compiled universal binary for Mac OS X 10.4.4 and higher is available at CRAN. This is distributed as a disk image containing the installer. In addition to the graphical interface version, a command line version (particularly useful for batch operations) can be run as the command `R`. A screenshot of the graphical interface can be found in Figure B.2. The GUI includes a mechanism to save and load the history of commands from within an interactive session. More information on Macintosh specific issues can be found in the CRAN *R for Mac OS X FAQ* (`http://cran.r-project.org/bin/macosx/RMacOSX-FAQ.html`).

B.1.3 Installation under Linux

Pre-compiled distributions of R binaries are available for the Debian, Redhat (Fedora), Suse and Ubuntu Linux, and detailed information on installation can be found at CRAN. There is no built-in graphical user interface for Linux (but see the R Commander project [24]).

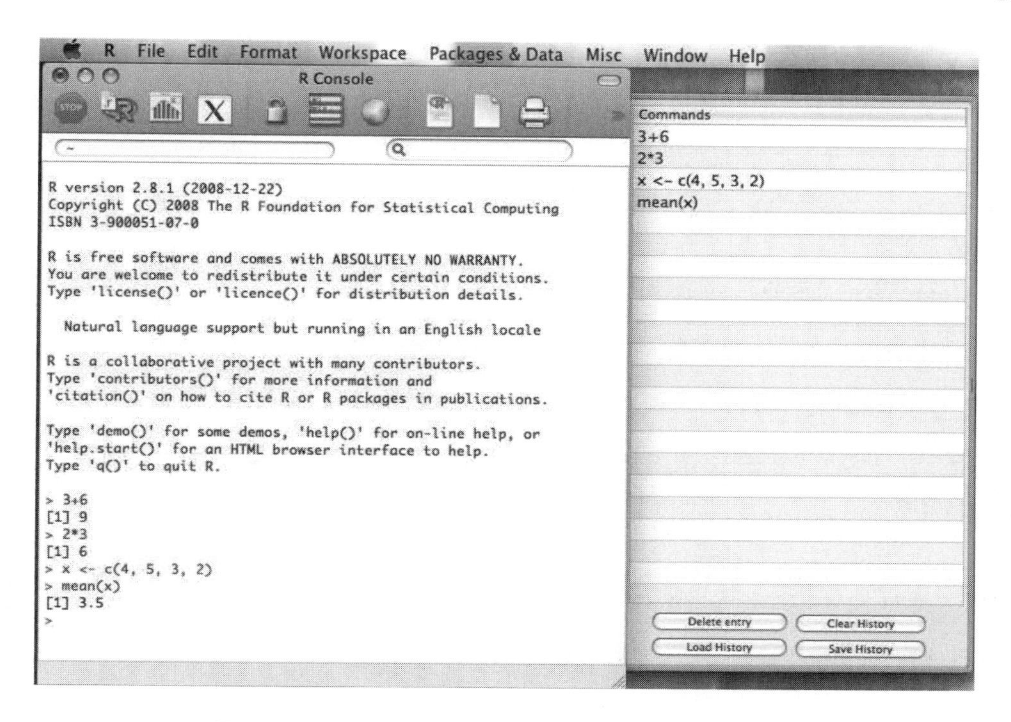

Figure B.2: R Mac OS X graphical user interface

B.2 Running R and sample session

Once installation is complete, the recommended next step for a new user would be to start R and run a sample session. An example from the command line interface within Mac OS X is given in Figure B.2.

The '>' character is the prompt, and commands are executed once the user presses the RETURN key. R can be used as a calculator (as seen from the first two commands on lines 1-4). New variables can be created (as on lines 5 and 8) using the assignment operator <-. If a command generates output (as on lines 6-7 and 11-12), then it is printed on the screen, preceded by a number indicating place in the vector (this is particularly useful if output is longer than one line, e.g., lines 24-25). A dataframe (akin to a dataset in SAS, here assigned the name `ds`) is read into R on line 15, then summary statistics are calculated (lines 22-23) and individual observations are displayed (lines 23-25). The `$` operator allows direct access to objects within a dataframe. Alternatively the `attach()` command can be used to make objects within a dataset available in the global workspace from that point forward.

Unlike SAS, R is case sensitive:

```
> x <- 1:3
> X <- seq(2, 4)
> x
[1] 1 2 3
> X
[1] 2 3 4
```

A very comprehensive sample session in R can be found in the Appendix A of *An Introduction to R* [94] (`http://cran.r-project.org/doc/manuals/R-intro.pdf`). New users to R will find it helpful to run the commands from that sample session .

```
% R
R version 2.8.1 (2008-12-22)
Copyright (C) 2008 The R Foundation for Statistical Computing
ISBN 3-900051-07-0

R is free software and comes with ABSOLUTELY NO WARRANTY.
You are welcome to redistribute it under certain conditions.
Type 'license()' or 'licence()' for distribution details.

  Natural language support but running in an English locale

R is a collaborative project with many contributors.
Type 'contributors()' for more information and
'citation()' on how to cite R or R packages in publications.

Type 'demo()' for some demos, 'help()' for on-line help, or
'help.start()' for an HTML browser interface to help.
Type 'q()' to quit R.

 1   > 3+6
 2   [1] 9
 3   > 2*3
 4   [1] 6
 5   > x <- c(4, 5, 3, 2)
 6   > x
 7   [1] 4 5 3 2
 8   > y <- seq(1, 4)
 9   > y
10   [1] 1 2 3 4
11   > mean(x)
12   [1] 3.5
13   > sd(y)
14   [1] 1.290994
15   > ds <- read.csv("http://www.math.smith.edu/sasr/datasets/help.csv")
16   > mean(ds$age)
17   [1] 35.65342
18   > mean(age)
19   Error in mean(age) : object "age" not found
20   > attach(ds)
21   > mean(age)
22   [1] 35.65342
23   > age[1:30]
24   [1] 37 37 26 39 32 47 49 28 50 39 34 58 53 58 60 36 28 35 29 27 27
25   [22] 41 33 34 31 39 48 34 32 35
26   > detach(ds)
27   > q()
28   Save workspace image? [y/n/c]: n
```

Figure B.3: Sample session in R

B.2.1 Replicating examples from the book and sourcing commands

To help facilitate reproducibility, R commands can be bundled into a plain text file, called a "script" file, which can be executed using the `source()` command. The optional argument `echo=TRUE` for the `source()` command can be set to display each command and its output. The book website cited above includes the R source code for the examples. The sample session in Figure B.2 can be executed by running:

```
> source("http://www.math.smith.edu/sasr/examples/sampsess.R", echo=TRUE)
```

while most of the examples at the end of each chapter can be executed by running:

```
> source("http://www.math.smith.edu/sasr/examples/chapterX.R", echo=TRUE)
```

where `X` is replaced by the desired chapter number. Many add-on packages need to be installed prior to running the examples (see B.6.1). To facilitate this process, we have created a script file to load them in one step:

```
> source("http://www.math.smith.edu/sasr/examples/install.R", echo=TRUE)
```

If these libraries are not installed (B.6.1), the example files at the end of the chapters will generate error messages.

B.2.2 Batch mode

In addition, R can be run in batch (non-interactive) mode from a command line interface:

```
% R CMD BATCH file.R
```

This will run the commands contained within `file.R` and put all output into `file.Rout`.

Special note for Windows users: to use R in batch mode, you will need to include R.exe in your path. In Windows XP, this can be accomplished as follows, assuming the default installation directory set up for R version 2.8.1. For other versions or non-default installations, the appropriate directory needs to be specified in the last step.

1. Right-click on "My Computer"
2. Click "Properties"
3. Select "Advanced" tab
4. Press "Environment Variables" button
5. Click "Path" (to highlight it)
6. Add `c:\program files\R\R-2.8.1\bin`

Once this is set up, the previously described `R CMD BATCH` syntax will work.

B.3 Learning R and getting help

As mentioned previously, an excellent starting point with R can be found in the *Introduction to R*, available from CRAN.

The system features extensive on-line documentation, though as with SAS, these can sometimes be challenging to comprehend. Each command in R has an associated help file that describes usage, lists arguments, provides details of actions, references, lists other related functions, and includes examples of its use. The help system is invoked using the command:

```
> ?function

or

> help(function)
```

where `function` is the name of the function of interest. As an example, the help file for the `mean()` function is accessed by the command `help(mean)`. The output from this command is provided in Figure B.3. It describes the `mean()` function as a generic function for the (trimmed) arithmetic mean, with arguments `x` (an R object), `trim` (the fraction of observations to trim, default = 0, trim = 0.5 is equivalent to the median), and `na.rm` (should missing values be deleted, default is `na.rm=F`). The function is described as returning a vector with the appropriate mean applied column by column. Related functions include `weighted.mean()` and `mean.POSIXct()`. Examples of many functions are available using the `example()` function:

```
> example(mean)
mean> x <- c(0:10, 50)
mean> xm <- mean(x)
mean> c(xm, mean(x, trim = 0.10))
[1] 8.75 5.50

mean> mean(USArrests, trim = 0.2)
  Murder  Assault UrbanPop     Rape
    7.42   167.60    66.20    20.16
```

Other useful resources are `help.start()`, which provides a set of online manuals and `help.search()`, which can be used to look up entries by description. The `apropos()` command returns any functions in the current search list that match a given pattern (which facilitates searching for a function based on what it does, as opposed to its name).

Other resources for help available from CRAN include the *Introduction to R* (described earlier) and the R-help mailing list (see also section B.7, support). New users are also encouraged to read the R FAQ (frequently asked questions) list.

B.4 Fundamental structures: objects, classes, and related concepts

Here we provide a brief introduction to R data structures. The *Introduction to R* (discussed in section B.2) provides more comprehensive coverage.

B.4.1 Objects and vectors

Almost everything in R is an object, which may be initially disconcerting to a new user. An object is simply something that R can operate on. Common objects include vectors, matrices, arrays, factors (see 1.4.12), dataframes (akin to datasets in SAS), lists, and functions.

The basic variable structure is a vector. Vectors can be created using the `<-` assignment operator (which assigns the evaluated expression on the right hand side of the operator to the object on the left hand side). For example:

```
> x <- c(5, 7, 9, 13, -4, 8)
```

creates a vector of length 6 using the `c()` function to concatenate scalars. Another assignment operator is `=`, which is generally recommended for the specification of options for

```
mean                   package:base                  R Documentation

Arithmetic Mean

Description:
     Generic function for the (trimmed) arithmetic mean.

Usage:
     mean(x, ...)

     ## Default S3 method:
     mean(x, trim = 0, na.rm = FALSE, ...)

Arguments:
       x: An R object.  Currently there are methods for numeric data
          frames, numeric vectors and dates.  A complex vector is
          allowed for 'trim = 0', only.

    trim: the fraction (0 to 0.5) of observations to be trimmed from
          each end of 'x' before the mean is computed. Values of trim
          outside that range are taken as the nearest endpoint.

   na.rm: a logical value indicating whether 'NA' values should be
          stripped before the computation proceeds.

     ...: further arguments passed to or from other methods.

Value:
     For a data frame, a named vector with the appropriate method being
     applied column by column.

     If 'trim' is zero (the default), the arithmetic mean of the values
     in 'x' is computed, as a numeric or complex vector of length one.
     If 'x' is not logical (coerced to numeric), integer, numeric or
     complex, 'NA' is returned, with a warning.

     If 'trim' is non-zero, a symmetrically trimmed mean is computed
     with a fraction of 'trim' observations deleted from each end
     before the mean is computed.

References:
     Becker, R. A., Chambers, J. M. and Wilks, A. R. (1988) _The New S
     Language_. Wadsworth & Brooks/Cole.

See Also:
     'weighted.mean', 'mean.POSIXct'

Examples:
     x <- c(0:10, 50)
     xm <- mean(x)
     c(xm, mean(x, trim = 0.10))
     mean(USArrests, trim = 0.2)
```

Figure B.4: Documentation on the `mean()` function

functions rather than assignment. Other assignment operators exist, as well as the `assign()` function (see section 1.11.5 or `help("<-")` for more information).

B.4.2 Indexing

Since vector operations are so common in R, it is important to be able to access (or index) elements within these vectors. Many different ways of indexing vectors are available. Here, we introduce several of these, using the above example. The command `x[2]` would return the second element of `x` (the scalar 7), and `x[c(2,4)]` would return the vector (7,13). The expressions `x[c(T,T,T,T,T,F)]`, `x[1:5]` and `x[-6]` (all elements except the 6th) would all return a vector consisting of the first five elements in `x`. Knowledge and basic comfort with these approaches to vector indexing is important to effective use of R.

Operations should be carried out wherever possible in a vector fashion (this is different from SAS, where data manipulation operations are typically carried out an observation at a time). For example, the expression:

```
> x>8
[1] FALSE FALSE  TRUE  TRUE FALSE FALSE
```

demonstrates the use of comparison operators. Only the third and fourth elements of `x` are greater than 8. The function returns a logical value of either `TRUE` or `FALSE`. A count of elements meeting the condition can be generated using the `sum()` function:

```
sum(x>8)
[1] 2
```

The code to create a vector of values greater than 8 is given by:

```
> largerthan8 <- x[x>8]
> largerthan8
[1] 9 13
```

in which `x[x>8]` can be interpreted as "the elements of x for which x is greater than 8." This is a difficult construction for some new users. Examples of its application in the book can be found in sections 1.4.14 and 1.13.2.

Other comparison operators include `==` (equal), `>=` (greater than or equal), `<=` (less than or equal and `!=` (not equal). Care needs to be taken in the comparison using `==` if non-integer values are present (see 1.8.5).

B.4.3 Operators

There are many operators defined in R to carry out a variety of tasks. Many of these were demonstrated in the sample section (assignment, arithmetic) and above examples (comparison). Arithmetic operations include `+`, `-`, `*`, `/`, `^` (exponentiation), `%%` (modulus), and `&/&` (integer division). More information about operators can be found using the help system (e.g., `?"+"`). Background information on other operators and precedence rules can be found using `help(Syntax)`.

R supports Boolean operations (OR, AND, NOT, and XOR) using the `|`, `&`, `!` operators and the `xor()` function, respectively.

B.4.4 Matrices

Matrices are rectangular objects with two dimensions. We can create a 2×3 matrix, display it, and test for its type with the commands:

```
> A <- matrix(x, 2, 3)
> A
     [,1] [,2] [,3]
[1,]    5    9   -4
[2,]    7   13    8
> # is A a matrix?
> is.matrix(A)
[1] TRUE
> is.vector(A)
[1] FALSE
> is.matrix(x)
[1] FALSE
```

Note that comments are supported within R (any input given after a # character is ignored).

Indexing for matrices is done in a similar fashion as for vectors, albeit with a second dimension (denoted by a comma):

```
> A[2,3]
[1] 8
> A[,1]
[1] 5 7
> A[1,]
[1]  5  9 -4
```

B.4.5 Dataframes

The main way to access data with R is through a dataframe, which is more general than a matrix. This rectangular object, similar to a dataset in SAS, can be thought of as a matrix with columns of vectors of different types (as opposed to a matrix, which consists of vectors of the same type). The functions `data.frame()`, `read.csv()`, (see section 1.1.4), and `read.table()` (see 1.1.2) return dataframe objects. A simple dataframe can be created using the `data.frame()` command. Access to sub-elements is achieved using the `$` operator as shown below (see also `help(Extract)`).

In addition, operations can be performed by column (e.g., calculation of sample statistics):

```
> y <- rep(11, length(x))
> y
[1] 11 11 11 11 11 11
> ds <- data.frame(x, y)
> ds
   x  y
1  5 11
2  7 11
3  9 11
4 13 11
5 -4 11
6  8 11
```

```
> is.data.frame(ds)
[1] TRUE
> ds$x[3]
[1] 9
> mean(ds)
        x         y
 6.333333 11.000000
> sd(ds)
       x        y
5.715476 0.000000
```

Note that use of `data.frame()` differs from the use of `cbind()`, which yields a matrix object:

```
> y <- rep(11, length(x))
> y
[1] 11 11 11 11 11 11
> newmat <- cbind(x, y)
> newmat
      x  y
[1,]  5 11
[2,]  7 11
[3,]  9 11
[4,] 13 11
[5,] -4 11
[6,]  8 11
> is.data.frame(newmat)
[1] FALSE
> is.matrix(newmat)
[1] TRUE
```

Dataframes can be conceived as the equivalent of datasets in SAS. They can be created from matrices using `as.data.frame()`, while matrices can be constructed using `as.matrix()`.

Dataframes can be attached using the `attach(ds)` command (see 1.3.1). After this command, individual columns can be referenced directly (i.e. `x` instead of `ds$x`). By default, the dataframe is second in the search path (after the local workspace and any previously loaded packages or dataframes). Users are cautioned that if there is a variable `x` in the local workspace, this will be referenced instead of `ds$x`, even if `attach(ds)` has been run. Name conflicts of this type are a common problem and care should be taken to avoid them.

The `search()` function lists attached packages and objects. To avoid cluttering the name-space, the command `detach(ds)` should be used once the dataframe is no longer needed. The `with()` and `within()` commands (see 1.3.1) can also be used to simplify reference to an object within a dataframe without attaching.

Sometimes a package (section B.6.1) will define a function (section B.5) with the same name as an existing function. If this occurs, packages can be detached using the syntax `detach("package:PKGNAME")`, where `PKGNAME` is the name of the package (see 4.6.5).

The names of all variables within a given dataset (or more generally for sub-objects within an object) are provided by the `names()` command. The names of all objects defined within an R session can be generated using the `objects()` and `ls()` commands, which return a vector of character strings.

The `print()` and `summary()` functions can be used to display brief or more extensive descriptions, respectively, of an object. Running `print(object)` at the command line is equivalent to just entering the name of the object, i.e. `object`.

B.4.6 Attributes and classes

Objects have a set of associated attributes (such as names of variables, dimensions, or classes) which can be displayed or sometimes changed. While a powerful concept, this can often be initially confusing. For example, we can find the dimension of the matrix defined earlier:

```
> attributes(A)
$dim
[1] 2 3
```

Other types of objects within R include lists (ordered objects that are not necessarily rectangular), regression models (objects of class `lm`), and formulas (e.g., `y ~ x1 + x2`). Examples of the use of formulas can be found in sections 2.4.1 and 3.1.1.

Many objects within R have an associated Class attribute, which cause that object to inherit properties depending on the class. Many functions have special capabilities when operating on a particular class. For example, when `summary()` is applied to a `lm` object, the `summary.lm()` function is called, while `summary.aov()` is called when an `aov` object is given as argument. The `class()` function returns the classes to which an object belongs, while the `methods()` function displays all of the classes supported by a function (e.g., `methods(summary)`).

The `attributes()` command displays the attributes associated with an object, while the `typeof()` function provides information about the object (e.g., logical, integer, double, complex, character, and list).

The `options()` function in R can be used to change various default behaviors, for example, the default number of digits to display in output (`options(digits=n)` where `n` is the preferred number). The command `help(options)` lists all of the other settable options.

B.5 Built-in and user-defined functions

B.5.1 Calling functions

Fundamental actions within R are carried out by calling functions (either built-in or user-defined), as seen previously. Multiple arguments may be given, separated by commas. The function carries out operations using these arguments using a series of pre-defined expressions, then returns values (an object such as a vector or list) that are displayed (by default) or saved by assignment to an object.

As an example, the `quantile()` function takes a vector and returns the minimum, 25th percentile, median, 75th percentile and maximum, though if an optional vector of quantiles is given, those are calculated instead:

```
> vals <- rnorm(1000) # generate 1000 standard normals
> quantile(vals)
     0%     25%     50%     75%    100%
-3.1180 -0.6682  0.0180  0.6722  2.8629
> quantile(vals, c(.025, .975))
 2.5% 97.5%
-2.05  1.92
```

Return values can be saved for later use.

```
> res <- quantile(vals, c(.025, .975))
> res[1]
 2.5%
-2.05
```

Options are available for many functions. These are named arguments for the function, and are generally added after the other arguments, also separated by commas. The documentation specifies the default action if named arguments (options) are not specified. For the `quantile()` function, there is a `type()` option which allows specification of one of nine algorithms for calculating quantiles. Setting `type=3` specifies the "nearest even order statistic" option, which is the default for SAS:

```
res <- quantile(vals, c(.025, .975), type=3)
```

Some functions allow a variable number of arguments. An example is the `paste()` function (see usage in 1.4.5). The calling sequence is described in the documentation as:

```
paste(..., sep=" ", collapse=NULL)
```

To override the default behavior of a space being added between elements output by `paste()`, the user can specify a different value for `sep`.

B.5.2 Writing functions

One of the strengths of R is its extensibility, which is facilitated by its programming interface. A new function is defined by the syntax `function(arglist) body`. The `body` is made up of a series of R commands (or expressions). Here, we demonstrate a function to calculate the estimated confidence interval for a mean from section 2.1.7.

```
# calculate a t confidence interval for a mean
ci.calc <- function(x, ci.conf=.95) {
    sampsize <- length(x)
    tcrit <- qt(1-((1-ci.conf)/2), sampsize)
    mymean <- mean(x)
    mysd <- sd(x)
    return(list(civals=c(mymean-tcrit*mysd/sqrt(sampsize),
                mymean+tcrit*mysd/sqrt(sampsize)),
                ci.conf=ci.conf))
}
```

Here the appropriate quantile of the T distribution is calculated using the `qt()` function, and the appropriate confidence interval is calculated and returned as a list. The function is stored in the object `ci.calc`, which can then be run interactively.

```
> ci.calc(x)
$civals
[1]  0.6238723 12.0427943
$ci.conf
[1] 0.95
```

If only the lower confidence interval is needed, this can be saved as an object:

```
> lci <- ci.calc(x)$civals[1]
> lci
[1] 0.6238723
```

The default confidence level is 95%; this can be changed by specifying a different value:

```
> ci.calc(x, ci.conf=.90)
$civals
[1]  1.799246 10.867421

$ci.conf
[1] 0.9
```

This is equivalent to running `ci.calc(x, .90)`. Other sample R programs can be found in sections 1.4.17 and 2.6.4.

B.5.3 The apply family of functions

Operations within R are most efficiently carried out using vector or list operations rather than looping. The `apply()` function can be used to perform many actions that would be implemented within a data step (section A.4) within SAS. While somewhat subtle, the power of the vector language can be seen in this example. The `apply()` command is used to calculate column means or row means of the previously defined matrix in one fell swoop:

```
> A
     [,1] [,2] [,3]
[1,]    5    9   -4
[2,]    7   13    8
> apply(A, 2, mean)
[1]  6 11  2
> apply(A, 1, mean)
[1] 3.333333 9.333333
```

Option 2 specifies that the mean should be calculated for each column, while option 1 calculates the mean of each row. Here we see some of the flexibility of the system, as functions in R (such as `mean()`) are also objects that can be passed as arguments to functions.

Other related functions include `lapply()`, which is helpful in avoiding loops when using lists, `sapply()` (see 1.3.2), and `mapply()` to do the same for dataframes and matrices, respectively, and `tapply()` (see 2.1.2) to perform an action on subsets of an object.

B.6 Add-ons: libraries and packages

B.6.1 Introduction to libraries and packages

Additional functionality in R is added through packages, which consist of libraries of bundled functions, datasets, examples and help files that can be downloaded from CRAN. The function `install.packages()` or the windowing interface under *Packages and Data* must be used to download and install packages. The `library()` function can be used to load a previously installed library (that has been previously made available through use of the `install.packages()` function). As an example, to install and load the `Hmisc` package, two commands are needed:

```
install.packages("Hmisc")
library(Hmisc)
```

Once a package has been installed, it can be loaded whenever a new session of R is run by executing the function `library(libraryname)`.

If a package is not installed, running the `library()` command will yield an error. Here we try to load the `Zelig` package (which had not yet been installed):

```
> library(Zelig)
Error in library(Zelig) : there is no package called 'Zelig'
```

```
> install.packages("Zelig")
trying URL 'http://cran.stat.auckland.ac.nz/bin/macosx/universal/contrib/
   2.8/Zelig_3.3-1.tgz'
Content type 'application/x-tar' length 14634987 bytes (14.0 Mb)
opened URL
==================================================
downloaded 14.0 Mb
The downloaded packages are in
/var/folders/ZI/ZIno+Dwy900va3+TU/-Tmp-//RtmpTaETVO/downloaded_packages
> library(Zelig)
Loading required package: MASS
Loading required package: boot
##
##  Zelig (Version 3.3-1, built: 2008-06-13)
##  Please refer to http://gking.harvard.edu/zelig for full documentation
##  or help.zelig() for help with commands and models supported by Zelig.
##
```

A user can test whether a package is available by running `require(packagename)`; this will load the library if it is installed, and generate an error message if it is not. The `update.packages()` function should be run periodically to ensure that packages are up-to-date.

As of March 2009, there were 1,705 packages available from CRAN. While each of these has met a minimal standard for inclusion, it is important to keep in mind that packages within R are created by individuals or small groups, and not endorsed by the R core group. As a result, they do not necessarily undergo the same level of testing and quality assurance that the core R system does.

B.6.2 CRAN task views

A very useful resource for finding packages are the *Task Views* on CRAN (`http://cran.r-project.org/web/views`). These are listings of notable packages within a particular application area (such as multivariate statistics, psychometrics, or survival analysis).

B.6.3 Installed libraries and packages

Running the command `library(help="libraryname")` will display information about an installed package. Entries in the book that utilize packages include a line specifying how to access that library (e.g., `library(foreign)`).

As of January 2009, the R distribution comes with the following packages:

base Base R functions

datasets Base R datasets

grDevices Graphics devices for base and grid graphics

graphics R functions for base graphics

grid A rewrite of the graphics layout capabilities, plus some support for interaction

methods Formally defined methods and classes for R objects, plus other programming tools

splines Regression spline functions and classes

stats R statistical functions

stats4 Statistical functions using S4 classes

tcltk Interface and language bindings to Tcl/Tk GUI elements

tools Tools for package development and administration

utils R utility functions

These are available without having to run the `library()` command and are effectively part of R.

B.6.4 Packages referenced in this book

Other packages utilized in the book include:

boot Bootstrap R (S-Plus) functions (Canty) [8]

circular Circular statistics [51]

coda Output analysis and diagnostics for MCMC [61]

coin Conditional inference procedures in a permutation test framework [37]

ellipse Functions for drawing ellipses and ellipse-like confidence regions [55]

elrm Exact logistic regression via MCMC [103]

epitools Epidemiology tools [4]

foreign Read data stored by Minitab, S, SAS, SPSS, Stata, Systat, dBase, ... [62]

gam Generalized additive models [30]

gee Generalized estimation equation solver [86]

ggplot2 An implementation of the Grammar of Graphics [100]

gmodels Various R programming tools for model fitting [97]

gtools Various R programming tools [96]

Hmisc Harrell miscellaneous [29]

irr Various coefficients of inter-rater reliability and agreement [25]

lattice Lattice graphics [73]

lme4 Linear mixed-effects models using S4 classes [5]

MCMCpack Markov chain Monte Carlo (MCMC) package [52]

mice Multivariate imputation by chained equations [93]

multcomp Simultaneous inference in general parametric models [36]

multilevel Multilevel functions [6]

nlme Linear and nonlinear mixed effects models [60]

plotrix Various plotting functions [44]

prettyR Pretty descriptive stats [45]

pscl Political science computational laboratory, Stanford University [104]

quantreg Quantile regression [42]

reshape Flexibly reshape data [99]

ROCR Visualizing the performance of scoring classifiers [81]

rpart Recursive partitioning [84]

survey Analysis of complex survey samples [50]

survival Survival analysis, including penalised likelihood [85]

vcd Visualizing categorical data [54]

VGAM Vector generalized linear and additive models [102]

XML Tools for parsing and generating XML within R and S-Plus [83]

Zelig Everyone's statistical software: an easy-to-use program that can estimate, and help interpret the results of, an enormous range of statistical models [39]

These must be downloaded, installed, and loaded prior to use (see `install.packages()`, `require()` and `library()`). To facilitate this process, we have created a script file to load these in one step (see B.2.1).

B.6.5 Datasets available with R

A number of data sets are available within the `datasets` package. The `data()` function lists these, while the optional `package` option can be used to specify datasets from within a specific package.

B.7 Support and bugs

Since R is a free software project written by volunteers, there are no paid support options available directly from the R Foundation. However, extensive resources are available to help users.

In addition to the manuals, FAQ's, newsletter, wiki, task views and books listed on the `www.r-project.org` web page, there are a number of mailing lists that exist to help answer questions. Because of the volume of postings, it is important to carefully read the posting guide at `http://www.r-project.org/posting-guide.html` prior to submitting a question. These guidelines are intended to help leverage the value of the list, to avoid embarrassment, and to optimize the allocation of limited resources to technical issues.

As in any general purpose statistical software package, bugs exist. More information about the process of determining whether and how to report a problem can be found using `help(bug.report()` (please also review the R FAQ).

Appendix C

The HELP study dataset

C.1 Background on the HELP study

Data from the HELP (Health Evaluation and Linkage to Primary Care) study are used to illustrate many of the entries in R and SAS. The HELP study was a clinical trial for adult inpatients recruited from a detoxification unit. Patients with no primary care physician were randomized to receive a multidisciplinary assessment and a brief motivational intervention or usual care, with the goal of linking them to primary medical care. Funding for the HELP study was provided by the National Institute on Alcohol Abuse and Alcoholism (R01-AA10870, Samet PI) and National Institute on Drug Abuse (R01-DA10019, Samet PI).

Eligible subjects were adults, who spoke Spanish or English, reported alcohol, heroin or cocaine as their first or second drug of choice, resided in proximity to the primary care clinic to which they would be referred or were homeless. Patients with established primary care relationships they planned to continue, significant dementia, specific plans to leave the Boston area that would prevent research participation, failure to provide contact information for tracking purposes, or pregnancy were excluded.

Subjects were interviewed at baseline during their detoxification stay and follow-up interviews were undertaken every 6 months for 2 years. A variety of continuous, count, discrete, and survival time predictors and outcomes were collected at each of these five occasions.

The details of the randomized trial along with the results from a series of additional analyses have been published [71, 66, 35, 47, 41, 70, 69, 79, 43, 101].

C.2 Roadmap to analyses of the HELP dataset

Table C.1 summarizes the analyses illustrated using the HELP dataset. These analyses are intended to help illustrate the methods described in the book. Interested readers are encouraged to review the published data from the HELP study for substantive analyses.

Table C.1: Analyses undertaken using the HELP dataset

Description	section (page)
Data input and output	1.13.1 (p.51)
Summarize data contents	1.13.1 (p.51)
Data display	1.13.2 (p.54)

Derived variables and data manipulation	1.13.3 (p.55)
Sorting and subsetting	1.13.4 (p.61)
Summary statistics	2.6.1 (p.78)
Exploratory data analysis	2.6.1 (p.78)
Bivariate relationship	2.6.2 (p.80)
Contingency tables	2.6.3 (p.82)
Two-sample tests	2.6.4 (p.85)
Survival analysis (logrank test)	2.6.5 (p.91)
Scatterplot with smooth fit	3.7.1 (p.111)
Linear regression with interaction	3.7.2 (p.113)
Regression diagnostics	3.7.3 (p.116)
Fitting stratified regression models	3.7.4 (p.119)
Two-way analysis of variance (ANOVA)	3.7.5 (p.120)
Multiple comparisons	3.7.6 (p.126)
Contrasts	3.7.7 (p.128)
Logistic regression	4.6.1 (p.146)
Poisson regression	4.6.2 (p.150)
Zero-inflated Poisson regression	4.6.3 (p.152)
Negative binomial regression	4.6.4 (p.154)
Quantile regression	4.6.5 (p.155)
Ordinal logit	4.6.6 (p.156)
Multinomial logit	4.6.7 (p.157)
Generalized additive model	4.6.8 (p.159)
Reshaping datasets	4.6.9 (p.160)
General linear model for correlated data	4.6.10 (p.164)
Random effects model	4.6.11 (p.166)
Generalized estimating equations model	4.6.12 (p.171)
Generalized linear mixed model	4.6.13 (p.172)
Proportional hazards regression model	4.6.14 (p.173)
Scatterplot with multiple y axes	5.6.1 (p.207)
Conditioning plot	5.6.2 (p.208)
Kaplan–Meier plot	5.6.3 (p.209)
ROC curve	5.6.4 (p.211)
Pairs plot	5.6.5 (p.213)
Visualize correlation matrix	5.6.6 (p.214)
Multiple imputation	6.5 (p.228)
Bayesian Poisson regression	6.6 (p.231)
Cronbach α	6.7.1 (p.233)
Factor analysis	6.7.2 (p.234)
Recursive partitioning	6.7.3 (p.237)
Linear discriminant analysis	6.7.4 (p.238)
Hierarchical clustering	6.7.5 (p.240)

C.3 Detailed description of the dataset

The Institutional Review Board of Boston University Medical Center approved all aspects of the study, including the creation of the de-identified dataset. Additional privacy protection was secured by the issuance of a Certificate of Confidentiality by the Department of Health and Human Services.

A de-identified dataset containing the variables utilized in the end of chapter examples is available for download at the book website:
`http://www.math.smith.edu/sasr/datasets/help.csv`.

Variables included in the HELP dataset are described in Table C.2. A copy of the study instruments can be found at: `http://www.math.smith.edu/help`.

Table C.2: Annotated description of variables in the HELP dataset

VARIABLE	DESCRIPTION	VALUES	NOTE
a15a	number of nights in overnight shelter in past 6 months	0–180	see also homeless
a15b	number of nights on the street in past 6 months	0–180	see also homeless
age	age at baseline (in years)	19–60	
anysubstatus	use of any substance post-detox	0=no, 1=yes	see also daysanysub
cesd*	Center for Epidemiologic Studies Depression scale	0–60	see also f1a–f1t
d1	how many times hospitalized for medical problems (lifetime)	0–100	
daysanysub	time (in days) to first use of any substance post-detox	0–268	see also anysubstatus
daysdrink	time (in days) to first alcoholic drink post-detox	0–270	see also drinkstatus
dayslink	time (in days) to linkage to primary care	0–456	see also linkstatus
drinkstatus	use of alcohol post-detox	0=no, 1=yes	see also daysdrink
drugrisk*	Risk-Assessment Battery (RAB) drug risk score	0–21	see also sexrisk
e2b*	number of times in past 6 months entered a detox program	1–21	
f1a	I was bothered by things that usually don't bother me	0–3#	
f1b	I did not feel like eating; my appetite was poor	0–3#	
f1c	I felt that I could not shake off the blues even with help from my family or friends	0–3#	
f1d	I felt that I was just as good as other people	0–3#	
f1e	I had trouble keeping my mind on what I was doing	0–3#	
f1f	I felt depressed	0–3#	
f1g	I felt that everything I did was an effort	0–3#	
f1h	I felt hopeful about the future	0–3#	
f1i	I thought my life had been a failure	0–3#	

f1j	I felt fearful	0–3#	
f1k	My sleep was restless	0–3#	
f1l	I was happy	0–3#	
f1m	I talked less than usual	0–3#	
f1n	I felt lonely	0–3#	
f1o	People were unfriendly	0–3#	
f1p	I enjoyed life	0–3#	
f1q	I had crying spells	0–3#	
f1r	I felt sad	0–3#	
f1s	I felt that people dislike me	0–3#	
f1t	I could not get going	0–3#	
female	gender of respondent	0=male, 1=female	
g1b*	experienced serious thoughts of suicide (last 30 days)	0=no, 1=yes	
homeless*	1 or more nights on the street or shelter in past 6 months	0=no, 1=yes	see also a15a and a15b
i1*	average number of drinks (standard units) consumed per day (in the past 30 days)	0–142	see also i2
i2	maximum number of drinks (standard units) consumed per day (in the past 30 days)	0–184	see also i1
id	random subject identifier	1–470	
indtot*	Index of Drug Abuse Consequences (InDuc) total score	4–45	
linkstatus	post-detox linkage to primary care	0=no, 1=yes	see also dayslink
mcs*	SF-36 Mental Composite Score	7-62	see also pcs
pcrec*	number of primary care visits in past 6 months	0–2	see also linkstatus, not observed at baseline
pcs*	SF-36 Mental Composite Score	14-75	see also mcs
pss_fr	perceived social supports (friends)	0–14	see also dayslink
satreat	any BSAS substance abuse treatment at baseline	0=no, 1=yes	
sexrisk*	Risk-Assessment Battery (RAB) drug risk score	0–21	see also drugrisk
substance	primary substance of abuse	alcohol, cocaine or heroin	
treat	randomization group	0=usual care, 1=HELP clinic	

Notes: Observed range is provided (at baseline) for continuous variables.
* denotes variables measured at baseline and followup (e.g., cesd is baseline measure, cesd1 is measure at 6 months, and cesd4 is measure at 24 months).
#: For each of the 20 items in HELP section F1 (CESD), respondents were asked to indicate

how often they behaved this way during the past week (0 = rarely or none of the time, less than 1 day; 1 = some or a little of the time, 1-2 days; 2 = occasionally or a moderate amount of time, 3-4 days; or 3 = most or all of the time, 5-7 days); items `f1d`, `f1h`, `f1l` and `f1p` were reverse coded.

Appendix D

References

[1] D. Adams. *The Hitchhiker's Guide to the Galaxy.* Pan Books, 1979.

[2] Alan Agresti. *Categorical Data Analysis.* John Wiley & Sons, 2002.

[3] J. Albert. *Bayesian Computation with R.* Springer, 2008.

[4] T. Aragon. *epitools: Epidemiology Tools*, 2008. R package version 0.5-2.

[5] D. Bates, M. Maechler, and B. Dai. *lme4: Linear mixed-effects models using S4 classes*, 2008. R package version 0.999375-28.

[6] P. Bliese. *multilevel: Multilevel Functions*, 2006. R package version 2.2.

[7] A. H. Bowker. Bowker's test for symmetry. *Journal of the American Statistical Association*, 43:572–574, 1948.

[8] A. Canty and B. Ripley. *boot: Bootstrap R (S-Plus) functions*, 2008. R package version 1.2-35.

[9] R. P. Cody and J. K. Smith. *Applied Statistics and the SAS Programming Language.* Prentice Hall, 1997.

[10] D. Collett. *Modelling Binary Data.* Chapman & Hall, 1991.

[11] D. Collett. *Modeling Survival Data in Medical Research (2nd edition).* Chapman & Hall/CRC, 2003.

[12] D. Collett. *Multivariate Statistical Methods: A Primer (3rd edition).* Chapman & Hall/CRC, 2004.

[13] L. M. Collins, J. L. Schafer, and C-M. Kam. A comparison of inclusive and restrictive strategies in modern missing data procedures. *Psychological Methods*, 6(4):330–351, 2001.

[14] R. D. Cook. *Residuals and Influence in Regression.* Chapman and Hall, 1982.

[15] L. D. Delwiche and S. J. Slaughter. *The Little SAS Book: a Primer (3rd edition).* SAS Publishing, 2003.

[16] A. J. Dobson and A. Barnett. *An Introduction to Generalized Linear Models (3rd edition).* Chapman & Hall/CRC, 2008.

[17] W. D. Dupont and W. D. Plummer. Density distribution sunflower plots. *Journal of Statistical Software*, 8:1–11, 2003.

[18] B. Efron and R. J. Tibshirani. *An Introduction to the Bootstrap*. Chapman & Hall, New York, 1993.

[19] J. J. Faraway. *Linear Models With R*. Chapman & Hall/CRC, 2004.

[20] J. J. Faraway. *Extending the Linear model with R: Generalized Linear, Mixed Effects and Nonparametric Regression Models*. Chapman & Hall/CRC, 2005.

[21] N. I. Fisher. *Statistical Analysis of Circular Data*. Cambridge University Press, 1996.

[22] G. M. Fitzmaurice, N. M. Laird, and J. H. Ware. *Applied Longitudinal Analysis*. Wiley, 2004.

[23] T. R. Fleming and D. P. Harrington. *Counting Processes and Survival Analysis*. John Wiley & Sons, 1991.

[24] J. Fox. The R Commander: a basic graphical user interface to R. *Journal of Statistical Software*, 14(9), 2005.

[25] M. Gamer, J. Lemon, and I. Fellows. *irr: Various Coefficients of Interrater Reliability and Agreement*, 2007. R package version 0.70.

[26] A. Gelman, J. B. Carlin, H. S. Stern, and D. B. Rubin. *Bayesian Data Analysis (2nd edition)*. Chapman and Hall, 2004.

[27] P. I. Good. *Permutation Tests: A Practical Guide to Resampling Methods for Testing Hypotheses*. Springer-Verlag, 1994.

[28] J. W. Hardin and J. M. Hilbe. *Generalized Estimating Equations*. Chapman & Hall/CRC, 2002.

[29] F. E. Harrell Jr. *Hmisc: Harrell Miscellaneous*, 2008. R package version 3.5-2.

[30] T. Hastie. *gam: Generalized Additive Models*, 2008. R package version 1.0.

[31] N. J. Horton, E. R. Brown, and L. Qian. Use of R as a toolbox for mathematical statistics exploration. *The American Statistician*, 58(4):343–357, 2004.

[32] N. J. Horton, E. Kim, and R. Saitz. A cautionary note regarding count models of alcohol consumption in randomized controlled trials. *BMC Medical Research Methodology*, 7(9), 2007.

[33] N. J. Horton and K. P. Kleinman. Much ado about nothing: A comparison of missing data methods and software to fit incomplete data regression models. *The American Statistician*, 61:79–90, 2007.

[34] N. J. Horton and S. R. Lipsitz. Multiple imputation in practice: comparison of software packages for regression models with missing variables. *The American Statistician*, 55(3):244–254, 2001.

[35] N. J. Horton, R. Saitz, N. M. Laird, and J. H. Samet. A method for modeling utilization data from multiple sources: application in a study of linkage to primary care. *Health Services and Outcomes Research Methodology*, 3:211–223, 2002.

[36] T. Hothorn, F. Bretz, and P. Westfall. Simultaneous inference in general parametric models. *Biometrical Journal*, 50(3):346–363, 2008.

[37] T. Hothorn, K. Hornik, M. A. van de Wiel, and A. Zeileis. Implementing a class of permutation tests: The coin package. *Journal of Statistical Software*, 28(8):1–23, 2008.

[38] R. Ihaka and R. Gentleman. R: A language for data analysis and graphics. *Journal of Computational and Graphical Statistics*, 5(3):299–314, 1996.

[39] K. Imai, G. King, and O. Lau. *Zelig: Everyone's Statistical Software*, 2008. R package version 3.3-1.

[40] S. R. Jammalamadaka and A. Sengupta. *Topics in Circular Statistics.* World Scientific, 2001.

[41] S. G. Kertesz, N. J. Horton, P. D. Friedmann, R. Saitz, and J. H. Samet. Slowing the revolving door: stabilization programs reduce homeless persons substance use after detoxification. *Journal of Substance Abuse Treatment*, 24:197–207, 2003.

[42] R. Koenker. *quantreg: Quantile Regression*, 2009. R package version 4.26.

[43] M. J. Larson, R. Saitz, N. J. Horton, C. Lloyd-Travaglini, and J. H. Samet. Emergency department and hospital utilization among alcohol and drug-dependent detoxification patients without primary medical care. *American Journal of Drug and Alcohol Abuse*, 32:435–452, 2006.

[44] J. Lemon, B. Bolker, S. Oom, E. Klein, B. Rowlingson, H. Wickham, A. Tyagi, O. Eterradossi, G. Grothendieck, M. Toews, J. Kane, M. Cheetham, R. Turner, C. Witthoft, J. Stander, and T. Petzoldt. *plotrix: Various plotting functions*, 2009. R package version 2.5-2.

[45] J. Lemon and P. Grosjean. *prettyR: Pretty descriptive stats*, 2009. R package version 1.4.

[46] K-Y Liang and S L Zeger. Longitudinal data analysis using generalized linear models. *Biometrika*, 73:13–22, 1986.

[47] J. Liebschutz, J. B. Savetsky, R. Saitz, N. J. Horton, C. Lloyd-Travaglini, and J. H. Samet. The relationship between sexual and physical abuse and substance abuse consequences. *Journal of Substance Abuse Treatment*, 22(3):121–128, 2002.

[48] S. R. Lipsitz, N. M. Laird, and D. P. Harrington. Maximum likelihood regression methods for paired binary data. *Statistics in Medicine*, 9:1517–1525, 1990.

[49] R. Littell, W. W. Stroup, and R. Freund. *SAS For Linear Models (4th edition).* SAS Publishing, 2002.

[50] T. Lumley. Analysis of complex survey samples. *Journal of Statistical Software*, 9(1), 2004.

[51] U. Lund and C. Agostinelli. *circular: Circular Statistics*, 2007. R package version 0.3-8.

[52] A. D. Martin, K. M. Quinn, and J. H. Park. *MCMCpack: Markov chain Monte Carlo (MCMC) Package*, 2008. R package version 0.9-5.

[53] P. McCullagh and J. A. Nelder. *Generalized Linear Models.* Chapman & Hall, 1989.

[54] D. Meyer, A. Zeileis, and K. Hornik. The strucplot framework: Visualizing multi-way contingency tables with vcd. *Journal of Statistical Software*, 17(3), 2006.

[55] D. Murdoch and E. D. Chow (porting to R by J. M. F. Celayeta). *ellipse: Functions for drawing ellipses and ellipse-like confidence regions*, 2007. R package version 0.3-5.

[56] P. Murrell. *R Graphics.* Chapman & Hall, 2005.

[57] P. Murrell. *Introduction to Data Technologies.* Chapman & Hall, 2009.

[58] N.J.D. Nagelkerke. A note on a general definition of the coefficient of determination. *Biometrika*, 78(3):691–692, 1991.

[59] National Institutes of Alcohol Abuse and Alcoholism, Bethesda, Maryland. *Helping Patients who Drink Too Much*, 2005.

[60] J. Pinheiro, D. Bates, S. DebRoy, D. Sarkar, and the R Core team. *nlme: Linear and Nonlinear Mixed Effects Models*, 2008. R package version 3.1-90.

[61] M. Plummer, N. Best, K. Cowles, and K. Vines. *coda: Output analysis and diagnostics for MCMC*, 2009. R package version 0.13-4.

[62] R-core members, S. DebRoy, R. Bivand, et al. *foreign: Read Data Stored by Minitab, S, SAS, SPSS, Stata, Systat, dBase, ...*, 2009. R package version 0.8-33.

[63] R Development Core Team. *R: A Language and Environment for Statistical Computing.* R Foundation for Statistical Computing, Vienna, Austria, 2008. ISBN 3-900051-07-0.

[64] T. E. Raghunathan, J. M. Lepkowski, J. van Hoewyk, and P. Solenberger. A multivariate technique for multiply imputing missing values using a sequence of regression models. *Survey Methodology*, 27(1):85–95, 2001.

[65] T E Raghunathan, P W Solenberger, and J V Hoewyk. IVEware: imputation and variance estimation software. *http://www.isr.umich.edu/src/smp/ive, accessed August 10, 2006*, 2006.

[66] V. W. Rees, R. Saitz, N. J. Horton, and J. H. Samet. Association of alcohol consumption with HIV sex and drug risk behaviors among drug users. *Journal of Substance Abuse Treatment*, 21(3):129–134, 2001.

[67] M. L. Rizzo. *Statistical Computing with R.* Chapman & Hall/CRC, 2007.

[68] D. B. Rubin. Multiple imputation after 18+ years. *Journal of the American Statistical Association*, 91:473–489, 1996.

[69] R. Saitz, N. J. Horton, M. J. Larson, M. Winter, and J. H. Samet. Primary medical care and reductions in addiction severity: a prospective cohort study. *Addiction*, 100(1):70–78, 2005.

[70] R. Saitz, M. J. Larson, N. J. Horton, M. Winter, and J. H. Samet. Linkage with primary medical care in a prospective cohort of adults with addictions in inpatient detoxification: room for improvement. *Health Services Research*, 39(3):587–606, 2004.

[71] J. H. Samet, M. J. Larson, N. J. Horton, K. Doyle, M. Winter, and R. Saitz. Linking alcohol and drug dependent adults to primary medical care: A randomized controlled trial of a multidisciplinary health intervention in a detoxification unit. *Addiction*, 98(4):509–516, 2003.

[72] J-M. Sarabia, E. Castillo, and D. J. Slottje. An ordered family of Lorenz curves. *Journal of Econometrics*, 91:43–60, 1999.

[73] D. Sarkar. *lattice: Lattice Graphics*, 2008. R package version 0.17-20.

[74] D. Sarkar. *Lattice: Multivariate Data Visualization With R.* Springer, 2008.

[75] C-E. Särndal, B. Swensson, and J. Wretman. *Model Assisted Survey Sampling.* Springer-Verlag, New York, 1992.

[76] J. L. Schafer. *Analysis of Incomplete Multivariate Data.* Chapman & Hall, 1997.

[77] R. L. Schwart, T. Phoenix, and b. d. foy. *Learning Perl (5th edition).* O'Reilly and Associates, 2008.

[78] G. A. F. Seber and C. J. Wild. *Nonlinear Regression.* Wiley, 1989.

[79] C. W. Shanahan, A. Lincoln, N. J. Horton, R. Saitz, M. J. Larson, and J. H. Samet. Relationship of depressive symptoms and mental health functioning to repeat detoxification. *Journal of Substance Abuse Treatment*, 29:117–123, 2005.

[80] T. Sing, O. Sander, N. Beerenwinkel, and T. Lengauer. ROCR: visualizing classifier performance in R. *Bioinformatics*, 21(20):3940–3941, 2005.

[81] T. Sing, O. Sander, N. Beerenwinkel, and T. Lengauer. *ROCR: Visualizing the performance of scoring classifiers*, 2007. R package version 1.0-2.

[82] B. G. Tabachnick and L. S. Fidell. *Using Multivariate Statistics (5th edition).* Allyn & Bacon, 2007.

[83] D. Temple Lang (duncan@wald.ucdavis.edu). *XML: Tools for parsing and generating XML within R and S-Plus*, 2009. R package version 2.3-0.

[84] T. M. Therneau and B. Atkinson (R port by B. Ripley). *rpart: Recursive Partitioning*, 2008. R package version 3.1-42.

[85] T. M. Therneau (ported by T. Lumley). *survival: Survival analysis, including penalised likelihood*, 2008. R package version 2.34-1.

[86] V. J. Carey. Ported to R by T. Lumley (versions 3.13, 4.4), and B. Ripley (version 4.13). *gee: Generalized Estimation Equation solver*, 2007. R package version 4.13-13.

[87] E. R. Tufte. *Envisioning Information.* Graphics Press, 1990.

[88] E. R. Tufte. *Visual Explanations: Images and Quantities, Evidence and Narrative.* Graphics Press, 1997.

[89] E. R. Tufte. *Visual Display of Quantitative Information (second edition).* Graphics Press, 2001.

[90] E. R. Tufte. *Beautiful Evidence.* Graphics Press, 2006.

[91] J. W. Tukey. *Exploratory Data Analysis.* Addison Wesley, 1977.

[92] S. van Buuren, H. C. Boshuizen, and D. L. Knook. Multiple imputation of missing blood pressure covariates in survival analysis. *Statistics in Medicine*, 18:681–694, 1999.

[93] S. van Buuren and C.G.M. Oudshoorn. *mice: Multivariate Imputation by Chained Equations*, 2007. R package version 1.16.

[94] W. N. Venables, D. M. Smith, and the R Development Core Team. An introduction to R: Notes on R: A programming environment for data analysis and graphics, version 2.8.0. *http://cran.r-project.org/doc/manuals/R-intro.pdf, accessed January 1, 2009*, 2009.

[95] J. Verzani. *Using R For Introductory Statistics*. Chapman & Hall/CRC, 2005.

[96] G. R. Warnes (includes R source code and/or documentation contributed by B. Bolker and T. Lumley). *gtools: Various R programming tools*, 2008. R package version 2.5.0.

[97] G. R. Warnes (includes R source code and/or documentation contributed by B. Bolker and T. Lumley and and R. C. Johnson). *gmodels: Various R programming tools for model fitting*, 2007. R package version 2.14.1.

[98] B. West, K. B. Welch, and A. T. Galecki. *Linear Mixed Models: A Practical Guide Using Statistical Software*. Chapman & Hall/CRC, 2006.

[99] H. Wickham. Reshaping data with the reshape package. *Journal of Statistical Software*, 21(12), 2007.

[100] H. Wickham. *ggplot2: An implementation of the Grammar of Graphics*, 2009. R package version 0.8.2.

[101] J. D. Wines, R. Saitz, N. J. Horton, C. Lloyd-Travaglini, and J. H. Samet. Overdose after detoxification: a prospective study. *Drug and Alcohol Dependence*, 89:161–169, 2007.

[102] T. W. Yee. *VGAM: Vector Generalized Linear and Additive Models*, 2009. R package version 0.7-8.

[103] D. Zamar, B. McNeney, and J. Graham. elrm: Software implementing exact-like inference for logistic regression models. *Journal of Statistical Software*, 21(3):1–18, 9 2007.

[104] A. Zeileis, C. Kleiber, and S. Jackman. Regression models for count data in R. *Journal of Statistical Software*, 27(8), 2008.

Appendix E

Indices

Separate indices are provided for subject (concept or task), SAS command, and R command. References to the HELP examples are denoted in *italics*.

Subject index

References to the HELP examples are denoted in *italics*.

SAS index

References to the HELP examples are denoted in *italics*.

R index

References to the HELP examples are denoted in *italics*.